T0075823

IRON, STONE AND STEAM

IRON, STONE AND STEAM

BRUNEL'S RAILWAY EMPIRE

TIM BRYAN

AMBERLEY

The Brunel Institute, housed at Brunel's SS *Great Britain* in Bristol is a collaboration of the SS Great Britain Trust and the University of Bristol.

The Institute houses the National Brunel Collection, which is the most important collection of Brunel-related material in the world and includes letters, calculation books, sketch books, plans, diaries, paintings, photographs, drawings and objects.

More details about the collection and access to the Institute's online database can be found at: https://www.ssgreatbritain.org/collections-and-research/

First published 2023

Amberley Publishing
The Hill, Stroud
Gloucestershire, GL5 4EP

www.amberley-books.com

British Library Cataloguing in Publication Data.
A catalogue record for this book is available from the British Library.

ISBN 978 1 3981 1269 8 (hardback)
ISBN 978 1 3981 1270 4 (ebook)

1 2 3 4 5 6 7 8 9 10

Typeset in 11pt on 14pt Sabon.
Typesetting by SJmagic DESIGN SERVICES, India.
Printed in the UK.

Contents

Author's Note and Acknowledgements

Although this book attempts to describe Brunel's exploits as a railway engineer in a broadly chronological manner, as the reader will discover, a number of the railways described were built concurrently in a flurry of activity across the West Country, Wales and the Midlands, and as a result some are dealt with in separate chapters for clarity. The first half of the book concentrates largely on the survey, promotion and construction of the Great Western Railway and details of surveys and the parliamentary process, for example, are not repeated for other Brunel railways described in later chapters.

It became apparent in the course of my research that there was a huge amount of historic material available to tell the story. It was inevitable therefore that to keep the book to a manageable size some railways are mentioned only briefly, so apologies are offered to those whose favourite Brunel projects are given less space than they might deserve. The bibliography contains detailed histories of some of these lines where appropriate and other detailed references and information are listed in the Endnotes.

For clarity, I have generally referred to Isambard Kingdom Brunel as 'Brunel' in the text, only using 'Isambard' where both he and his father are mentioned. I have also retained the original nineteenth-century Welsh place names rather than their modern equivalents.

Having written a book on the life of Isambard Kingdom Brunel more than twenty years ago I had always wanted to revisit the subject and having returned to my home city of Bristol in 2019 as Director of the Brunel Institute at the SS Great Britain Trust, it seemed a perfect opportunity to write again about the work of one of Britain's most iconic engineers, this time focussing on his railway works. The Institute contains the National Brunel collection, which includes the archive previously held at the University of Bristol Library and I have been fortunate enough to be able to draw on primary and secondary material from the Institute collection and I am grateful to SS Great Britain Trust Director and CEO Dr Matthew Tanner and Trustees for their support and for permission to reproduce historic material and images.

I would particularly like to thank the Brunel Institute collections team of Joanna Mathers, Mollie Bowen, Victoria Haddock and Abi Mann for their patience and help during my research, and the contribution of the Brunel Institute volunteer team, in particular Jan Bowen and Andrew Ballinger. I am also grateful to James Boyd, Head of Research at the Brunel Institute, for his help, and the support of colleagues in the wider SS Great Britain Trust team, particularly Phil Smith, Angela Spreadbury, Nerys Watts, Rachel Roberts, Nicola Grahamslaw, Carol Griffin, Paul Chibeba, Liam Tolhurst, and Joe Teanby. Special thanks are due to Marnie Rees for her invaluable IT assistance.

Much of the other primary research was undertaken at the National Archives, the British Library, Search Engine at the National Railway Museum York, and the reference collections of the Bristol and Swindon Central Libraries. I am very grateful to the staff in these organisations for their assistance. Elaine Arthurs and Felicity Jones at STEAM – Museum of the Great Western Railway in Swindon kindly made available material from the library and archive there and also assisted with images. Peter Rance, Frank Dumbleton and other members of the Great Western Trust team at Didcot were also very generous and helpful, providing me with access to their archive. Thanks are also due to Carol Morgan, archivist at the Institution of Civil Engineers, Nick Booth and the library team at the Ironbridge Gorge Museum Trust in Coalbrookdale, and Dr Clare George, archivist, Senate House Library, University College London.

Many other individuals provided assistance and support: Dr Helen Doe and Christian Wolmar provided helpful advice in the early stages of planning. Peri and James Martin and Hugh James kindly showed me around Brunel's historic dock at Briton Ferry and Simon Gyde of Network Rail showed me around Temple Meads. I am grateful to Mr Howard Beard for permission to quote from the diary of Charles Richardson. Thanks are also due to colleagues at the Railway Heritage Trust and members of the National Brunel Network including Julia Elton, David Greenfield, Stephen Jones and Ann Middleton. Other colleagues and friends past and present including The Canon Reverend Brian Arman, Ralph Buckland, Mike Fenton, Steve Gregory, Jack Hayward, Rosa Matheson helped directly and indirectly. Adrian Vaughan shared his insights and knowledge, and Gregory Spawton provided inspiration with music from the Underfall Yard. Extra special thanks are due to Steph Gillett who kindly read and commented on the draft and provided moral support.

Finally, this book could not have been completed without the support of my family and I'd like to thank Jack and Katie and Kate for their encouragement. My wife Ann put up with my absence while I researched and wrote, and she read and edited the text and provided the love and support needed to complete it.

Swindon, 2023

Introduction

The monumental towers of the Royal Albert Bridge at Saltash spanning the River Tamar between Devon and Cornwall bear the name of their engineer Isambard Kingdom Brunel and commemorate one of the last and perhaps most enduring achievements of his career. By the time the bridge was finally completed in 1859, Brunel, worn out and ill, was too unwell to attend its grand opening ceremony, but was able to see his final creation as he was hauled across it on a railway wagon a few weeks later.

Work on the bridge and Brunel's third steamship, the SS *Great Eastern*, then anchored on the Thames after a difficult construction and launch, had dominated the last years of his life and marked the culmination of his career as a civil, mechanical and railway engineer. Much has been written about Brunel since his death and the charismatic and original engineer will always be remembered for his three steamships, the SS *Great Western*, *Great Britain* and *Great Eastern*, his early designs for the Clifton Suspension Bridge and his part in helping his father Marc build the Thames Tunnel, along with many other engineering projects. It was his role as a railway engineer, though, that cemented his reputation as one of the preeminent engineers of the nineteenth century, and his most significant legacy is his work on the railways. The soaring roof at Paddington station and the graceful arches of Maidenhead Bridge, his railway village and works at Swindon, the daring tunnel at Box and his pioneering terminus at Bristol Temple Meads along with many other bridges, stations and structures remain as a lasting testament to his talents as a railway engineer.

Brunel's twenty-six-year career as a civil and railway engineer spanned an astonishing period of social, political and industrial change and his life straddled the old world of the Regency era and the brave new world of Victorian Britain; 'Nobody...will doubt for a moment that we are living at a period of the most wonderful transition,'[1] noted Prince Albert, for whom Brunel's Saltash bridge had been named. It is clear that the engineer's railway adventures coincided with dramatic changes in railway development as well as profound technological, political and social change that would characterise the period more generally. When Brunel was appointed as the engineer to the Great Western Railway in 1833, railways were still very much in their infancy and to most of the British population still a real novelty; the first inter-city line, Stephenson's Liverpool & Manchester, had only been open for three years, and other major new railways were either still being planned or in the course of construction. Brunel arrived on the scene at a key moment when large-scale railway development was about to explode into existence, dramatically changing the landscape, economy and way of life for British people.

Although Brunel was an integral part of that railway revolution, as this book will show, his ideas and experiments would result in railways that did not exactly fit the accepted evolutionary pattern being developed and practised by Stephenson, Locke and others, and it could be argued that his idealistic concept for railways provided a new and alternative paradigm to the accepted way of doing things. Brunel described his Great Western Railway as 'the finest work in England' even before it had been built, and when it was finally completed, it incorporated many innovations which gave his conception a distinctly different look to any other railway then planned or built. There is no doubt that Brunel's work as a railway engineer demonstrated not only an astonishing level of technical skill, but also a singular design vision, demonstrating a diverse range of influences, ideas and styles, melding art, science and engineering in his designs. Brunel's rejection of established practice and custom and his propensity for doing the opposite of what might be expected did not always succeed, however, often making him unpopular with Directors, shareholders and the press.[2]

As a result, the book will highlight not only some of Brunel's career-defining railway achievements and successes but will also aim to balance these triumphs with some analysis of his failures. Brunel was, the Bristol chronicler John Latimer thought, 'an inexperienced theorist, enamoured with novelty, prone to seek for difficulties rather than evade them, and utterly indifferent to the outlay which his recklessness entailed on his employers'. His 'pet crotchet' Latimer continued, was the broad gauge, the bold and ultimately doomed idea of introducing a track gauge of 7 ft ¼ in for his railways rather than the 4 ft 8½ in 'standard gauge' favoured by the 'sober-minded, practical and economical engineers of the North'.[3] Brunel's unconventional baulk road track, and the initial batch of 'freak' locomotives built for the GWR to his eccentric specifications also provided his opponents with ample ammunition to criticise his work even before the completion of the original line from Bristol to London. Perhaps the most infamous chapter in his career was the unsuccessful atmospheric railway system installed on the South Devon Railway, which ultimately cost its shareholders more than £400,000 and dealt a serious but not a fatal blow to Brunel's reputation. Despite this seemingly severe dent to his professional standing, Brunel – 'the Napoleon of engineers, thinking more of glory than profits' – ploughed on, taking this setback and others like it in his stride, simply moving on to the next challenge.

Brunel's railway empire would eventually grow to cover more than twelve hundred miles of lines, beginning with the original Great Western Railway from Bristol to London. It has often erroneously been said that Brunel built the GWR, but as many have pointed out, his railways were the result of the 'labour of body and mind'[4] of the thousands of navvies who literally carved out his railways from the landscape of England and Wales. From the start, Brunel's vision encompassed an interconnected network of broad-gauge railways that would eventually run west via the Bristol & Exeter, South Devon and Cornwall and West Cornwall railways and North from Swindon to Gloucester and Cheltenham via the Cheltenham & Great Western Union Railway. Trains running into the Principality ran from Gloucester westwards on the South Wales Railway, and his work extended to other enterprises in the Forest of Dean and the South Wales valleys. Further East, trains would be able to run through Oxford, and then on to the Midlands via the Oxford & Rugby and the Oxford, Wolverhampton &

Worcester line. South of the original GWR main line, railways like the Berks & Hants, and Wilts, Somerset and Weymouth probed deep into southern England; these railways, along with other smaller branches and connecting lines made up a substantial part of what would eventually become the Great Western Railway, most being absorbed into the larger company before the creation of the 'Big Four' railways in 1923. Brunel's influence also extended beyond Britain, and as a consulting engineer he was responsible for railways in Ireland, Italy and India, particularly in the years after the 'Railway Mania', when contracts at home began to dry up somewhat.

More than 180 years after their creation, many of the lines, stations and structures created by Brunel remain as a testament to his determination and imagination, but despite the sometimes monumental scale of his railways, this book does not just focus on the tangible outcomes of his work. Instead, the story of Brunel's railway empire is also a very human story, encompassing not only his sometimes complicated and contradictory character but also his influences, motivations and working methods, and his interactions with staff, railway directors, shareholders, contractors and workmen who turned his grand conceptions into reality. Underlying all was a relentless drive to succeed, perhaps typical of the times. D. R. Burgess reflected that 'pushing the limits of technology reinforced the qualities Victorians most esteemed in themselves: perseverance, genius and service to the common good.'[5]

While Brunel's public persona was one of a brilliant, cultured man full of ambition, confidence, wit and self-assurance, his early diaries in particular reveal a vein of pessimism and self doubt rarely shown to the wider world, fuelled perhaps by the insecurity of his early life, by the experiences of his father and the need for financial security and public recognition. He was perfectly aware of his 'self conceit and love of glory', writing that he wished to be 'the first engineer and example to all others'[6] but later recording that his success might not last, and that 'bad weather must surely come'. Brunel's drive to succeed manifested itself in an almost superhuman workload, made far worse by his reluctance to delegate and his obsessive eye for detail that on occasions threatened the success of his projects in itself, and made the lives of his assistants and the Directors of his railways difficult. On the eve of the opening of the first section of the GWR in 1838 he wrote to his old friend Thomas Guppy that he was 'nervous, anxious and unhappy – in fact blue devilish'. There were 'an infinite number of things requiring attention and thought,' he complained, 'all in arrears and I am quite incapable of getting through them', concluding that he needed a dose of salts'.[7] The complex and flawed personality of the engineer meant that while he could be charm itself to those he liked and respected, he could treat others badly, like the contractors building his railways with whom he often had a difficult relationship, particularly when they failed to live up to his notoriously high standards.

Brunel's office diaries reveal a man constantly on the move, particularly in the early days of his career, and while he found time to marry in 1836, it seems likely that his wife Mary saw him only intermittently during the busiest professional periods of his life. As he surveyed the routes of his new railway lines, travelling from his office in London across the West Country and Wales, he criss-crossed the country first on horseback and later in his own horse-drawn 'travelling hearse' amply equipped as a mobile office and a place to rest as he moved from one place to another. As his rail network grew, he then made considerable use of his own lines, often leaving on very early trains from Paddington to the West Country, the Midlands and Wales. Did Brunel have a gold season ticket one wonders, or was his stovepipe hat, slightly rumpled clothes and ever-present cigar enough for railway staff to recognise him? His physical appearance did not always impress, and a humorous thumbnail of Brunel published a year before his death in 1859 by a correspondent in the *Manchester Guardian* described 'a short man of about five and forty…in his second best hat, with a shocking bad suit'.[8]

As well as learning something of Brunel's character, we shall also meet a gallery of colourful characters who played their part in the Brunel story; these include the early Bristol Directors like Thomas Guppy and Nicholas Roch who saw something in the brash young man desperate to launch his career in the city after almost losing his life in the Thames Tunnel, and stalwart supporters like Charles Saunders, Secretary of the Great Western Railway, and George Henry Gibbs, one of the railway's Directors who remained loyal even when Brunel's constant quest for experiment and innovation put the whole future of the railway in peril. Through letters and diaries, we also meet some of Brunel's long-suffering staff of clerks, draughtsmen, resident engineers and surveyors who worked both in his Duke Street Office and around the vast railway network he built. Some like Daniel Gooch would become well known in their own right as the GWR grew, but others are now only remembered through surviving archive material that records their interactions with their sometimes irascible but always driven boss.

Finally, we consider the flamboyant engineer's dealings with a motley cast of other characters, most of whom were fundamentally opposed to Brunel and his railway projects; disgruntled shareholders, vitriolic newspapermen, quack scientists, venal landowners, unprincipled lawyers, opinionated members of parliament, rival engineers and railway managers, and many others who in the earliest days of railway development feared the coming of a phenomenon that would change their world for ever, but not, in their opinion, for the better.

Inevitably, the story of Isambard Kingdom Brunel and his railways is one that has been described before by his family, contemporaries and railway historians, biographers and engineers and it is impossible to tell that story without acknowledging the debt owed to them and the unnamed clerks, draughtsmen and writers who recorded the work and activities of Brunel and the companies he worked for. The thousands of documents, letters, drawings and sketches that survive in archives, libraries and museums ensure that there will always be more to discover and learn. We will remain curious and fascinated in equal measure about the continuing legacy of Brunel and his railways.

Prologue

The neat rows of terraced houses that included Britain Street in Portsmouth have long gone, the birthplace of one of Britain's most famous Victorian engineers marked today by a rather nondescript plaque in a quiet spot not far from the town's historic naval dockyard. Born on 9 April 1806, Isambard Kingdom Brunel was the third child of Marc Isambard and Sophia Kingdom Brunel, and he and his siblings, the five-year-old Sophia and two-year-old Emma, lived in a very modest house in Portsmouth where their father was employed by the Admiralty supervising the construction of a works to manufacture blocks for Navy ships.

Born in 1769 in Hacquville, a small town in Normandy, Marc Isambard Brunel had fled from France in 1793 as the turmoil of the revolution that had broken out four years earlier continued, his Royalist views making the guillotine a very real fate if he did not leave the country. Marc had met Sophia Kingdom, the sixteen-year-old daughter of a Plymouth naval contractor in Paris in 1792, and the couple had become engaged in the months before Marc was forced to travel to New York to evade the Republican terror. It would be six years before they would meet again, by which time Marc had begun to build a career as an architect and engineer in the United States. In the interim Sophia had been imprisoned by the French authorities and only returned to England in 1795 after much tribulation.

While in America, Marc Isambard Brunel had dined with Alexander Hamilton, and met another French émigré, M. Delagibarre, who told him that the British Royal Navy was encountering difficulties in manufacturing enough pulley blocks for their warships, a slow process that was for the most part still done by hand. Convinced he could transform this task through mechanisation, Marc obtained an introduction from Hamilton to the Admiralty, and set sail for England, arriving there in January 1799. He was also anxious to be with Sophia, and after a long-awaited reunion in March they were finally married in London on 1 November 1799.

Despite having little formal training, Marc was one of the most talented and inventive engineers of his generation and in the course of the next three years, he was able to convince the Admiralty that the machines he had designed to manufacture blocks would work and be a success. In this process he was assisted by Henry Maudsley, another engineer who would play a significant part in the careers of both Marc and his son in the future. It was Maudsley's models of the block-making machines that would help persuade Samuel Bentham, the Inspector-General of Navy Works, to accept Brunel's ideas, and in 1802 the Brunel family moved to Portsmouth as work to build and commission a block-making factory under Marc's supervision began there. The block-making machines proved to be a great success, with a manufacturing process originally requiring the work of sixty men being reduced to only six, but disputes with the Navy over what would now be called intellectual property rights and renumeration meant that Marc was eventually paid only £17,000 for his invention; a substantial sum, but not what the work was worth.

This disappointment meant that in 1808, the family were on the move again, as Marc sought to expand his engineering practice and improve his financial position. Marc, Sophia and their three children moved into a grander London house in Cheyne Walk near the Thames, and it was here that the young Isambard grew up while his father launched two new businesses, neither of which would ultimately prove financially successful. The first was a partnership in a sawmill and veneer works in Battersea that Marc Brunel thought, like his block-making process, could be improved by mechanisation. The second was a speculative venture which initially seemed to be a great success; Marc again applied new technology to the manufacture of a product that supported the war against France, this time to the making of good-quality boots for British soldiers fighting in the war with

Napoleon, which had been rumbling on since 1803. By 1812 his factory was turning out 400 pairs per day, but after Wellington's victory at Waterloo in 1815, with no large army to support the British government not only ended its order for boots but refused to pay for those it had already ordered, leaving Marc Brunel out of pocket. In a situation that would become familiar to the Brunel family, their financial situation was grim, and it became grimmer in August 1814 when the sawmill burned to the ground and it became clear that the partner Marc had trusted had been less than honest, leaving him with only £865 in the bank, not the £10,000 expected. Undeterred, Marc rebuilt the sawmill and relaunched the enterprise. Matters improved a little, but the elder Brunel was no businessman, and the family lived in a constant state of financial uncertainty and worry.

Whatever his financial shortcomings, Marc Brunel was undoubtedly a gifted engineer, and was determined that his son should follow him into the profession. As well as teaching his son mathematics from an early age, Marc also taught Isambard to sketch and draw, encouraging him to practise regularly. Despite the family's troubles, Isambard was sent to study at Dr Morell's boarding school in Hove, and in 1817 was packed off to France by his father to be tutored there by his nephew for a year or so. In 1820, Marc judged that the restoration of the French monarchy and the reestablishment of some calm meant he could send the young Isambard first to the College at Caen and then the Lycée Henri-Quatre in Paris, before spending time with the famous clockmaker Breguet. Isambard also studied at the Institution De M. Massin. The elder Brunel held the quality of French schooling in the highest regard and hoped that his son would also improve his knowledge of French, which had 'become rusty at school'. It seems, however, that his French had not improved enough by the summer of 1822, since he failed to pass the entrance examination to enter the prestigious Ecole Polytechnique, where competition for places was extremely fierce. If Isambard had managed to gain one of the 100 places at the Polytechnique just over two hundred years ago, the Brunel story might have been very different, but it was not to be, and on 21 August 1822 the sixteen-year-old Isambard Kingdom Brunel returned home to London after spending nearly three years in France.

While he had been away, Marc had suffered a final indignity when his bankers failed, and he had been declared insolvent. With no immediate way of settling his debts and the government still refusing to pay, he was sent to the Kings Bench, a debtors' prison, and languished there for almost three months. He and Sophia, who had loyally joined him in the prison, were only able to leave in July after he had threatened to take up an invitation from Tsar Alexander to work in Russia. This finally prompted a payment of £5,000 from the government but left him with little else to invest in speculative projects like the sawmill.

Although Isambard had been able to remain in France cared for by family and friends, he never forgot the dark times his parents had experienced, and the need for both fame and fortune drove the young engineer forward in the coming years. When he arrived back in London the young Brunel was put to work in his father's office at No. 29 Poultry, close to the Bank of England, where Marc had set up in the more secure role of a consulting engineer. Whatever doubts Sophia may have had about her son following her husband into the family business, the young Isambard seems not to have had any similar misgivings, having long been prepared for this life by his father. The variety of projects Marc was then working on provided the civil and mechanical engineering training that would be the foundations of the young engineer's future career. Within a year, Isambard would be helping his father design the Thames Tunnel, and by 1826 working in the dark, stinking and dangerous depths of the tunnel itself, an appointment that would ultimately end in near tragedy when the River Thames flooded the workings.

1

Bristol Connections

Late on the evening of Boxing Day 1835, the 29-year-old Isambard Kingdom Brunel took the opportunity to make a lengthy entry in his personal journal, at what he clearly judged to be a momentous point in his fledgling career. The entry, full of typical Brunel bravado, includes a commentary on a list of projects he was then working on and his thoughts about his future prospects.

The triumphant tone of the diary is a world away from the despair and frustration that seeped from the pages of his personal journal written in the wilderness years after being seriously injured in an accident in the Thames Tunnel on 12 January 1828. On that day the tunnel had flooded for the second time and when the river burst in, Brunel's leg had been trapped under a timber before a wave of filthy water swept him almost 600 feet along the unfinished tunnel. Brunel's battered body was then carried up the shaft at Rotherhithe. Suffering internal injuries, Brunel had a lucky escape, unlike the six workers who did not survive the disaster. Four months after his brush with death in the murky water of his father's tunnel, he laid in bed smoking a pipe, musing that while things were gloomy, he could not 'work myself up to be downhearted' but that ultimately he might be doomed to 'a sort of middle path' and that a 'mediocre success' as an engineer on £200 or £300 a year might be his fate.[1]

That Boxing Night in 1835, how much had changed in the intervening seven years is revealed. Isambard noted that he had not written in his journal for almost twelve months; six months earlier on 31 August, the bill authorising the Great Western Railway had finally been passed by Parliament following an incredibly intense period of activity when he had been engaged in the survey, design and promotion of the line and this achievement, along with the benefit of a £2,000 a year salary, marked a watershed for him. Though it was not this achievement alone: 'Everything I have been engaged in has been successful,' he wrote.

Brunel listed four other railway projects for which he was now engineer as a result of his work on the GWR, not all seemingly as important or interesting to him, even at this early stage. The Merthyr and Cardiff, later to be named the Taff Vale Railway, was dismissed with the offhand comment 'I care not however about it' and the Newbury Branch was 'a little go almost beneath my notice now'. The Cheltenham Railway was similarly of little interest and Brunel argued that he was only engaged in it 'because they can't do without me' and it was only the Bristol & Exeter Railway that seemed to have captured his imagination. Still yet to be authorised by Parliament, this railway according to Isambard had a survey 'done in grand style'. With relatively few engineering difficulties and a capital of £1,250,000 it probably posed less of a challenge than those he subsequently would face on the GWR and other railways in the coming years. As an afterthought, Brunel added that he 'forgot also the Bristol & Gloster Railway'; this ugly duckling line would nevertheless play an important part in his railway empire, then only a dream for the engineer.[2]

Adding in bridge and dock works, Brunel reflected that he had 'a pretty list of real profitable, sound professional jobs' adding that most had been largely unsought for by him, but 'given fairly by the respective parties'. Adding together all the capital invested in these projects, which amounted to more than £5 million, he boasted that he had made his fortune 'or rather the foundation of it'; this new-found prosperity meant that he could move from 53 Parliament Street to the rather grander surroundings of 18 Duke Street, where he could both live and have a larger office for his burgeoning workload.

As the pace of Brunel's working career accelerated, he never again found the time to properly record his personal thoughts, although as Angus Buchanan records, it may simply be that later diaries, unlike much else recording his life and work, simply did not survive. But as we begin to chronicle Isambard's career as a railway engineer in more detail, we need to explore what brought him from the disappointment and relative obscurity of 1828 to the stepping off point of 1835, where, as he noted, he had become the engineer to the 'finest work in England' and he could assert that he was 'now somebody'.

Brunel's recovery from the calamitous accident at the Thames Tunnel was slow. It was thought that convalescence at Brighton would be helpful, but it appears that the bright lights and sea air of the Regency resort were a step too far, and despite meeting what he recorded as 'pleasant company', after two weeks it seemed that his stay had proved too much for his still delicate constitution and after suffering a haemorrhage he was forced to return to his bed. His full recovery eventually took more than six months, and as work on the Thames Tunnel proceeded only fitfully, he had little to do initially. Brunel's son later recorded that he was 'without regular occupation' for much of the rest of 1828, although he was able to have a brief holiday in Plymouth and then France, returning to England in February 1829. His diary records that he filled his time with visits to distinguished scientists such as Babbage and Faraday and attending lectures at the Institution of Civil Engineers and the Royal Society. By the summer of 1829, Brunel was well enough to undertake further convalescence away from home, this time in the City of Bristol, a place that was destined to change his fortunes for good.

Brunel had described the intervening years between his accident and his appointment as engineer to the Great Western Railway as a stage in his life when he had been 'toiling most unprofitably at numerous things'.[3] It does seem that during this period he was able to earn a modest income but exactly how he did this is not entirely clear, as evidence for how he managed it is fragmentary and based on incomplete diary entries and accounts. Few of the projects he was then engaged on seem to have come to fruition after optimistic beginnings while others took far longer to complete than he would have expected. With the suspension of work on the Thames Tunnel, Brunel and his father continued to work on their experiments to create a 'Gaz' engine, a method of generating gas from ammonium carbonate and sulphuric acid using condensers, which they hoped would generate a source of energy that might replace steam, a source of power still in its infancy itself. Marc and

Isambard had already invested much time and money on the idea with little success and in April 1829 Brunel noted in his diary that he had spent a further six months experimenting on the Gaz engine to no avail. Some income was generated by a contract to implement drainage works at Tollesbury in Essex involving the installation of a steam pumping engine supplied by Maudsley Son & Field and a 'siphon' to be used on the sea wall. Early accounts and diaries reveal that the young engineer charged five guineas a day for his services with additional charges for travel and payment for the services of the assistants and draughtsmen he was able to employ at that time.[4]

When a further spell of convalescence was proposed, Brunel travelled to the more sedate surroundings of Bristol where he stayed in the fashionable suburb of Clifton. A few weeks later he learned of a competition to design a bridge over the Avon Gorge, funded by a bequest originally provided by a Bristol wine merchant, William Vick, as early as 1753. Vick's original £1,000 gift had been given to the Society of Merchant Venturers, an ancient Bristol guild, whose origins could be traced back to the sixteenth century. By 1829 the original bequest was worth almost £8,000 and a committee was set up to implement a scheme to design and build a bridge across the 700-foot-deep gorge. On 1 October 1829 a competition was announced, with a prize of 100 guineas for the successful designer. After surveying the site, Brunel initially submitted four designs, and these were shortlisted along with the work of four other engineers in November. Unable to make a decision, the judging committee enlisted the services of Thomas Telford, the designer of the Conway and Menai bridges, and the most respected civil engineer of the time. Telford 'reported unfavourably' on all the proposed designs arguing that none were suitable and was requested to prepare a design himself 'adapted to the boldness of the cliffs and the beauty of the surrounding scenery'.[5] Thus it was that Telford's own bridge plans were submitted with a bill to Parliament in 1830, but the enthusiasm for his clumsy Gothic design quickly faded and almost a year after the original competition, the bridge committee agreed to hold a second competition, this time with Telford submitting designs rather than judging them. Brunel provided new plans, but was disappointed again, his entry being placed second. Undeterred, he asked for a further interview with the judges and bridge trustees and was able not only to overturn their original decision but also to persuade them to adopt his proposal to decorate the bridge in an Egyptian style. Brunel and his father, who had already assisted him greatly in work on the bridge, approached fellow architect Auguste Pugin, who produced a design idea on the Egyptian theme which helped sway the judges and was later produced as a lithograph.[6] Writing to Benjamin Hawes, Brunel boasted that the idea had been 'quite extravagantly admired by all' by 'fifteen men who were quarrelling about the most ticklish subject – taste'.[7] Brunel's victory that day was just the first of many occasions where reluctant investors or Directors were persuaded to back his schemes by his charisma and oratory, not always with the best outcomes.

Having won over the Bridge committee with his designs and been appointed as engineer with a salary of 2,000 guineas a year Brunel might have expected his first major commission to proceed without too much difficulty. Writing to his brother-in-law Benjamin Hawes in March 1831 he thought the completion of the bridge would 'be a pleasant job, for the expense seems no object provided it is made grand'. On 18 June work did begin on preparing the ground on the Clifton side of the gorge, and a few days later, after a breakfast at the Bath Hotel, where Brunel was often to stay while on business in Bristol, a ceremony was held to mark the start of work. A band played, cannons were fired, and speeches made, including one by local landowner Sir Abraham Elton, who promised that the time would soon come when 'Brunel was recognised in the streets of every city…the man who reared that stupendous work, the ornament of Bristol and the wonder of the Age.'[8]

The optimism and public enthusiasm for the Clifton Bridge project did not last long, however, and Brunel's confidence was premature; it soon became apparent that despite the protracted recruitment process for the project and the grandiose fanfare just described, Brunel's hope that expense might be no object was misplaced and the bridge committee was nowhere near financially solvent enough to complete the project and had so far failed to raise enough money to bolster the original Vick bequest. The already shaky situation was then completely overtaken by events in October 1831 when Bristol was plunged into chaos as three days of riots sparked by the rejection of Lord Grey's Reform Bill by the House of Lords saw public buildings such as the Mansion House burned or ransacked, and hundreds of rioters roaming the streets. Discontent and social and political instability had gripped the country for some years, but there is little doubt that local frustration at the corruption and inefficiency of Bristol's corporation and public officials also played its part. Brunel enrolled as a special constable and as might be expected seems to have been in the thick of the action.[9]

As the fires were damped down and the process of rebuilding the shattered city began, it was natural that the idea of raising more money for the Clifton Bridge project was for the foreseeable future a distant prospect. Brunel was forced to return to working on a variety of new and existing projects; he continued his experiments on the Gaz engine when time permitted, but a month after the riots he was to be found in Sunderland meeting with local businessmen and merchants about a plan to build a new dock at Monkwearmouth. The long journey to the North East was worth it, as despite some reservations on Brunel's part about the committee, he was appointed as engineer for the dock scheme.[10] Initial plans were rejected by Parliament, but work on the dock finally began in 1835.

Despite the fact that the Thames Tunnel had been bricked up and all work had ceased in 1829, Isambard had continued to support his father Marc; there was little enthusiasm for investing in what the press had dubbed 'The Great Bore' and efforts to obtain loans from the government to complete the project were unsuccessful. To make matters worse, his father suffered a heart attack at the end of November 1831 and the younger Brunel recorded in his diary that the 'Tunnel is now I think DEAD.' Marc Isambard Brunel, having come so far, did not share his son's pessimism, and continued to lobby for the tunnel, although it would be another three years before he was finally able to find the financial backing he needed. When work on the tunnel began again four years later in January 1835, Isambard was completely absorbed in work on the Great Western Railway and Richard Beamish, Marc Brunel's former assistant, was instead appointed as resident engineer.

Brunel's Bristol connections provided a further opportunity in the summer of 1832. Brunel had met Nicholas Roch, who was a Bristol oil and leather manufacturer, at the Mansion House during the riots a year earlier as they rescued valuables from the damaged building during the disturbances, and he would also have also encountered him as one of the Trustees of the Clifton Bridge project. Impressed by the young engineer's talents and energy, and as a Director of the Bristol Dock Company, Roch was instrumental in Brunel obtaining a commission to report on possible improvements to Bristol's Floating Harbour. The company had been set up in 1802 to build and maintain the harbour, which had been designed by William Jessop to provide the city with a dock where ships could remain afloat whatever the state of the tide outside lock gates; with a tidal range of around 30 ft, the old port was difficult to operate and Jessop's new harbour provided over 80 acres of enclosed water, larger than the docks in Liverpool or London when built.

By the 1830s, silt and sewage were making the docks increasingly difficult to operate, and Brunel's report completed in August 1832 emphasised the need to increase the flow of water running through the harbour to prevent further silt and mud building up, and made a number of recommendations, including the rebuilding of a dam at Netham to raise its level by one foot to improve water flow, the building of a sluice at the Princes Street Bridge to help scour the docks, and the construction of a 'drag boat' to dredge the harbour itself.[11] Progress on implementing any of these innovations was slow, restricted by caution and lack of finance following the Bristol Riots and it was more than a year before the Dock Company commissioned some of the improvements, including the construction of a drag boat following an inspection of the harbour by Brunel in February 1833 when it had been completely drained of water.

On that visit to Bristol, Brunel met with Roch, who once again provided provided him with news of another opportunity, this time a proposal by the Bristol business community to promote and build a railway to link the city to London. Bristol's position as one of the most important and successful port cities in the country had declined as rivals such as Liverpool began to grow, and if it was to sustain the prosperity founded on the trade in tobacco, sugar and slaves it had developed in the 17th and 18th centuries, good communication with the capital was essential. While turnpike roads had improved significantly, a journey from Bristol to London could still take more than sixteen hours, with numerous changes of horses along the Great West Road. Goods haulage was slower and expensive; the Kennet and Avon Canal opened in 1810 provided the means to move freight between the two cities, but shortages of water, pilferage and bad weather were significant issues for those using it.

The opening of the Stockton & Darlington Railway in 1825 had in fact prompted the Directors of the Kennet & Avon Canal Company to send their own engineer, Mr Blackwell North, to see it and other new railways at work and 'to report his opinions and observations thereon'; but the growth of railways and the use of steam power was no mere curiosity.[12] By 1800 a large number of horse-drawn wagonways linking mines, canals and ports had already been well established and though George Stephenson's Stockton & Darlington line was not the first public railway, it was the first really successful railway to use steam locomotives to transport freight and passengers on wrought iron rails.

The business community in Bristol had also been aware of the potential of steam railways, with no fewer than three schemes being promoted in 1824, a year before the opening of Stephenson's line. One of these was a proposed line from Bristol to Bath and its prospectus boasted that the railway was not based on 'speculative or visionary assertions' but relied on facts collected by observing the practices of 'Rail-Roads' in the North of England.[13] These plans foundered when half the city's banks failed in a financial crash the same year, but two years later a Bristol chronicler recorded that 'The iron railways, as they are strangely denominated, are still going on with wonderful earnestness. I rather doubt their success, although I have a few shares myself.'[14]

The opening of the Liverpool & Manchester Railway in 1830 had by far the greatest impact locally and nationally. The railway was the world's first intercity line and brought together many features of what would become the modern railway; described as 'an idea in the right place at the right time'[15] the Liverpool & Manchester combined new, more robust and powerful locomotives and rails capable of handling heavier trains. Most importantly, the new railway company was operating a potentially profitable route that linked the growing port of Liverpool and the city of Manchester, which was rapidly becoming the largest and most important manufacturing centre in England. Already worried by the loss of business to Liverpool, the Bristol business community could not fail to see glowing reports of the success of the

Liverpool & Manchester and as a result two further schemes to build a railway from Bristol to London were promoted in 1832. A prospectus for the 'Bristol and London Railway' was published in May that year and boasted that 'the novel and extraordinary effects of railways mark an important era in our internal policy, and will most effectually contribute to revive the languishing commerce of the country.'[16] The proposed railway was promoted by the civil engineer William Brunton, who noted that when completed, the new line would allow passengers to complete the journey between the two cities in six hours. Confusingly, another prospectus, this time titled as the 'London to Bristol Railway' was published a month later from a different London address, but with text identical to the earlier document.[17] Both proposals shared the same route, the line running through Bath, Trowbridge, Devizes, Hungerford, Newbury, Reading and then on to London with a terminus planned close to Edgware Road and Oxford Street. The estimated cost of the new line was £2,500,000 and while neither prospectus appears to have inspired enough investors to fund the project, their demise was not the end of the story.

In the autumn of 1832, a group of four influential Bristol men, George Jones, John Harford, Thomas Guppy and William Tothill met in a small office in Temple Back, a street not far from the present-day Temple Meads station. The group were determined to make another attempt to promote a railway from Bristol to London; there was little doubt that they were public spirited enough to be convinced of the need to improve the financial standing of the city, but the fact that shares in the Liverpool & Manchester Railway were selling at double their original value, and those of the Stockton & Darlington Railway treble, must have also been a motivating factor.[18] After this informal meeting a larger, more formal committee of fifteen prominent members of the business community was constituted to develop the idea. The enlarged group included representatives of five of the most important business organisations then operating in the city, the Bristol Corporation, the Society of Merchant Venturers, The Bristol Dock Company, The Bristol Chamber of Commerce and the Bristol & Gloucester Railway.[19] Thomas Guppy, whom Brunel had met as early as 1829, was the only member of the original steering group not to join the full committee initially, although once financial support for the new railway was secured and committees of Directors from both Bristol and London were established to oversee the promotion and construction of the line, he would join the Bristol committee. Guppy was also a key figure in the establishment of the Great Western Steamship Company, which was responsible for the construction of Brunel's SS *Great Western* and SS *Great Britain*, as well as an investor in a number of other railway companies linked to the GWR.

The new steering committee met on 21 January 1833 and following confirmation that all the organisations represented were in favour of the scheme, it was agreed that 'the expediency of promoting the formation of a Rail Road from Bristol to London'[20] should be fully investigated. A circular was then sent to 'a large number of firms and individuals' covering agricultural, mercantile and manufacturing interests to canvass opinion on the possible benefits of a railway and it later emerged that 32 favourable responses from 'gentlemen of the highest respectability' had been received. With the support of such important interests within the city, the new scheme had a far better chance of success than the speculative projects previously mooted. The five bodies further agreed to fund the next stage of the project, a survey of the route of the new line. Nicholas Roch, who was on the committee as a representative of the Bristol Dock Company, was asked to advise on the appointment of an engineer to undertake the survey.

On 21 February, a month after the new committee had met, Roch saw Brunel in Bristol at the offices of Osborne, the solicitor acting for the new railway and discussed the project with him. Brunel reported in his diary Roch had 'informed me in his presence that a sub-committee was appointed for the purpose of receiving offers from me'.[21] The first mention in Isambard Kingdom Brunel's diary of what was to become the Great Western Railway appeared on that day as a marginal reference noted 'B.R.', initials that then referred to the 'Bristol Railway'. The appointment process for the new railway was not, however, straightforward; Brunel was to be in competition with Brunton, engineer to the abortive schemes already mentioned, and more importantly he was told that he would have to work with W.H. Townsend, a local engineer and surveyor, who had been responsible for the Bristol & Gloucestershire railway, a horse-drawn tramway that ran from coal mines at Coalpit Heath to

Cuckolds Pill on the River Avon in Bristol. The short nine-mile line had been the first railway in the city and Parliament had approved its construction in June 1828, with the object of 'making a cheaper and more expeditious conveyance for coal and stone to the City of Bristol' and for carrying other goods to and from the 'populous' mining districts it served.[22]

Already unhappy he was to be saddled with Townsend, Brunel then discovered that the appointment would be made to the engineer who provided the lowest cost estimate for the construction of the line, following an initial survey. Despite Roch recommending this course of action, Brunel told him that 'you are holding out a premium to the man who will make you the most flattering promises,' and argued that this arrangement was simply not acceptable to him, adding that he would not take the job on that basis, and that he would provide the committee with a route that was the best, but not necessarily the cheapest. One wonders quite what Roch must have thought of this, but he patiently asked Brunel to set down his views on paper for the consideration of the railway committee. This Brunel did in a six-page letter which began by stating that it would let them know 'the terms on which "we" would furnish a plan and section from actual and personal survey for a...line you may consider the most suitable'. The survey would enable the committee to select from 'the different practicable lines, one which offers most commercial advantages and of which you must be better judges than an engineer'. He also argued that although this would only be a preliminary survey, any work done could form part of a future Parliamentary survey, rather than answering the immediate purpose of 'laying a favourable line before the public to be afterwards useless'.[23] Brunel also offered the assistance of his father before noting that to complete the work within two months and prepare plans to which he would attach his name and 'for which my reputation would be answerable would require great exertion'. With regard to expenses, Isambard told the committee that both he and Townsend had agreed to 'merely seek our actual disbursements' and would accept a payment of no less than £500.

Having stated his case, the engineer then returned to London to attend the Annual General Meeting of the Thames Tunnel Company, returning by the night mail coach in the early hours of

6 March. The railway committee met that morning to consider the various survey proposals, and after a long debate Brunel was given the good news that he had been appointed as engineer to the new railway, along with Townsend as already intimated. His appointment was formally confirmed later that afternoon at the Council House, by the committee chairman, John Cave. The following day, Brunel dined with Roch and discovered that Cave, a Bristol banker and Master of the Society of Merchant Venturers, had voted against his appointment and as a result he had been selected as engineer to the new railway by the smallest margin – just one vote.

As he had done when bidding for the Clifton Bridge position, Brunel had gambled on his reputation and ability to influence and convince potential employers and investors in order to succeed, something he would subsequently do on numerous occasions on other projects. This uncompromising mixture of confidence and arrogance might be forgiven if Brunel had been an established engineer with successful projects under his belt, but as a relatively inexperienced 27-year-old, his determination to succeed on his own terms is remarkable. However close Isambard had come to failing just at the point he had been presented with his biggest opportunity, the job was his, and the appointment would provide a springboard for his future career and cement his reputation as one of the foremost railway engineers in Britain.

It is worth reflecting that for all his charisma and confidence Brunel was perhaps still fortunate to have been appointed in such circumstances, since on the face of it, he would have seemed a somewhat unlikely candidate for the job, given that he had very little direct experience of railway engineering in comparison to rivals like George and Robert Stephenson. It seems, therefore, that while prominent members of the Bristol business community were very anxious to promote and build a new railway to enhance the city's status on the national stage, they took a more far parochial approach in the recruitment of the engineer for the line and the initial proposal to appoint on the basis of a cheapest estimate seems also to reflect this lack of ambition.

There does not seem to be any evidence that the railway committee approached George or Robert Stephenson or advertised more widely for other potential candidates. Railway engineering was at that time still a relatively new field and as a result the pool of talent from which potential candidates for the job could be chosen was small; between 1825 and 1833 more than forty new railway schemes had been authorised by Parliament, with the Stephensons being responsible for over a quarter of these.

While many of the pioneering engineers of the early nineteenth century such as Telford, Maudsley, Naysmyth and George and Robert Stephenson had learnt their craft as artisans and mechanics who had little access to formal education, Brunel had a somewhat different apprenticeship. His early education in France from 1820 to 1822 had prepared him for a more practical grounding in civil and mechanical engineering in his father's office. As Angus Buchanan noted, while Isambard was not born with a silver spoon in his mouth, 'he could at least be regarded as having had a spanner in his hand from birth – or rather a draughtsman's eye and a hand for executing a fine drawing.'[24] The family connection cannot fail to have been an advantage for Brunel, and would certainly have been something the Bristol railway committee were aware of when they appointed him.

Brunel's apprenticeship with his father meant that he would have had access to the scientific and engineering community from an early age and as an inquisitive and ambitious young engineer would of course have closely followed the progress of railways as the new and rising technology of the age, through newspapers and journals of the time and through professional and personal contacts in London and elsewhere. Isambard had been elected to the Institution of Civil Engineers in 1829 and would have been able not only to consult the then modest library held by the Institute, but also attend some of the regular meetings where original communications (OCs) or manuscript papers from members were read. C.B. Vignoles produced a number of papers relating to locomotives and railways and by 1832 further papers on railways had begun to appear, including one by Combe on the Liverpool and Manchester Railway in 1831. Brunel was elected a Fellow of the Royal Society in 1830 and his early journals reveal that he attended meetings there, often with his father, mixing with many of the leading scientists and engineers of the time, although there is little evidence he was an active participant, or submitted papers to the Royal Society or any of the other professional bodies of which he was a member.[25]

No records appear to have survived of the contents of Brunel's technical library, but it is likely that he would have owned copies of the most up-to-date books on steam engines and railway technology then available, such as Thomas Tredgold's 1827 *The Steam Engine*, which the author described as 'an account of the invention and progressive improvement of the steam engine', whose power was 'gigantic, and involves so many new and important doctrines in mechanical science and practice'. The book covered the development of the steam engine and its science, technology and application in mines, agriculture, manufacturing, steam navigation and most importantly what Tredgold called 'land carriage by railways'.[26]

It also seems likely that Isambard would have talked a great deal about railway developments with his father Marc while living with his parents in London; in 1828, the elder Brunel had produced a schematic map titled 'Railways from London to Birmingham and to Bristol'.[27] The simple plan identified these three cities as hubs for railway lines from London to Birmingham, London to Bristol, and Bristol to Birmingham, with an additional central station at Oxford that would also have individual links to all three major destinations. There is no attempt to plot out the actual geographical route of each line, but the distances between each city are recorded, as are other potential stops such as Bath on the western route and towns such as Gloucester and Tewkesbury on the Bristol-Birmingham line. While Sir Marc's scheme may have appeared logical at the time, local, national and railway company politics would mean that it did not ultimately resemble the network his son would eventually create in the next twenty or so years, but the idea of a broader intercity railway network south of Birmingham may well have influenced Isambard's future thinking.

Contemporary records suggest that Brunel had visited few of the railways then in operation before his appointment as engineer to the GWR, although in a letter applying for a job as engineer to the Newcastle & Carlisle Railway on 19 December 1829 he appears to imply that this was not the case, since after noting that he had gained much experience as a civil engineer having been 'brought up under my father', Brunel wrote that he had devoted 'much attention to the subject of railways' and that he was 'well acquainted by personal inspection with visits of the principal works of this kind'.[28] The promotion of the Newcastle & Carlisle line, intended to link the North East of England with the Irish Sea, had been a complex and protracted process, hampered by opposition from landowners and various proposals for the route dating back to at least 1825; it was not until May 1829 that parliamentary approval for the railway was finally agreed.

Brunel had been suggested as a potential candidate for the job when it became apparent that the Directors of the Liverpool & Manchester Railway were unlikely to release George Stephenson for the position. Writing about the recruitment process on 23 December, one of the railway's promoters, James Lock, recorded that Lord Carlisle had a 'strong inclination for Brunel who he suggests is a clever young man'. Despite this recommendation and others from supporters like Lord Lonsdale, the Directors of the Company wrote to Brunel on 8 January 1830 to inform him that they had offered the post of engineer to Francis Giles, 'upon certain conditions' but that they had resolved that if these conditions were not accepted by Giles that the appointment would be offered to Brunel. This rebuff must have been a blow to Brunel's pride; most historians are agreed that Giles was an undistinguished canal engineer of limited talent who had famously said of George Stephenson that 'No engineer in his senses would go through Chat Moss if he wanted to make a railway from Liverpool to Manchester.' The process of subsequently completing the Newcastle & Carlisle was a slow and drawn-out process, with the line not completed until 1838, so Brunel may have been fortunate to have been rejected on this occasion.

A further unsuccessful footnote in Brunel's early involvement in railway projects before his eventual appointment as engineer to the GWR took place in November 1830, when he travelled to Birmingham to meet the promoters of a railway from Birmingham to Bristol. A provisional committee of Birmingham businessmen had, like their counterparts in Bristol, recognised the success of the Liverpool & Manchester and other new railways and were anxious to revive a scheme originally proposed in 1826. Brunel had been introduced to J.W. Whatley, the solicitor to the new company, by family friend Charles Babbage, and he travelled up to the Midlands in November 1830 to meet him. The interview seems to have gone well, and Brunel was further encouraged after dining with members of the provisional committee for the line and was subsequently appointed to undertake a survey of the route, revising original work done by Josiah Jessop five years earlier. In an article published in a Birmingham newspaper a month or so before, it had been reported that the new scheme, then titled the Birmingham, Worcester, Gloucester & Bristol Railway, would link the city with the ports of Gloucester and Bristol as well as connecting with Worcester and other centres of population, but it appears Brunel was instead instructed to survey a more direct route between Birmingham and Gloucester that avoided intermediate stops at places like Tewkesbury, Worcester or Droitwich, where high land values might increase costs.

This more commercial approach meant that Brunel's route ran east of the route now used by trains and ran through the Lickey Hills instead. No evidence of his original survey appears to have survived, but this route would have probably involved some steep gradients; in the event, Brunel's estimates were deemed too expensive by the company, and plans for the railway were abandoned. Brunel was paid £100 for his work, and when the project was revived some years later in 1835, a new westerly route engineered by Captain Moorsom was the one adopted.[29]

Brunel's diaries reveal that he had visited two of the most important railway projects in operation by 1831, both engineered by George Stephenson. Following the meeting with the promoters of a project to build a new dock at Monkwearmouth in late November 1831 mentioned earlier, Brunel went on to visit the Stockton & Darlington Railway where he inspected a suspension bridge carrying the railway over the River Tees – which he described as a 'wretched thing' – and then travelled on to Hartlepool, Beverley and Hull. From Hull he crossed the Pennines to Manchester to undertake a journey on the Liverpool & Manchester Railway, which at that point had only been running for just over a year. There are no previous mentions of railway travel in Brunel's diaries, so it is likely that this was his first trip on a train. It seems Brunel was not too impressed by the rough ride he had on a 2 hour 15 minute journey to Liverpool on 5 December 1831, writing in his notebook 'I record this specimen of the shaking on the Manchester Railway – the time is not too far off when we shall be able to take our coffee and write while going, noiseless and smoothly at 45 miles per hour – let me try.'[30]

Brunel tried to draw circles in his notebook with mixed success to judge how smooth the track was, and on one page he tried to write his signature twice at both top and bottom, along with the words 'Shaky Shaky'.[31] Within a few short years however, Brunel would be grappling with his own issues of rough track as teething troubles with his new GWR line led to a battle with shareholders and critics that almost cost him his job.

With his appointment as engineer to the new railway confirmed on 6 March 1833 and undoubtedly confident he could do the job, Brunel was nevertheless still required to work with W. H. Townsend, despite having written of him in his diary 'How the devil am I to get on with him tied to my neck.' Townsend did have a good reputation locally but appears to have accepted a more subservient role soon after work began on the survey, despite contractually being on an equal footing with Brunel. Both men had agreed to undertake the survey for the £500 noted in Brunel's letter to the committee and there was no time to lose, as a deadline of May had been set for the completion of the work. Energised by the prospects of creating the 'finest work in England', Brunel rapidly embarked on the survey of the railway and the contrast between the two men soon became apparent; beginning their work in Bristol, Townsend suggested that the new line should initially follow the route of his Bristol & Gloucester railway climbing out of the city to Mangotsfield through what Brunel called 'dreadfully hilly country' before descending again into Bath. Brunel, with the vision of a fast intercity line in mind, instead suggested a more direct route along the Avon valley, at the same time deciding that the Bristol terminus should be at Temple Meads.

To complete even this very preliminary survey involved traversing the whole of the proposed route on horseback and foot, agreeing a potential route, taking initial levels of the landscape and approaching landowners whose property might well be impacted by the new railway crossing their property. Brunel embarked on what author Adrian Vaughan called an 'itinerant horseback life', riding hundreds of miles as he scouted the best route for the railway. With Townsend working largely at the Bristol and Bath end of the proposed route, Brunel employed surveyor Thomas Hughes to help him, although he was not always as diligent as Isambard might have wished. Within months, two routes had been considered, a southerly line passing through Bradford-on-Avon, the Kennet valley, the Marlborough Downs and Vale of Pewsey and Newbury and Reading, and a more northerly route via Bath, Swindon, the Vale of the White Horse and the Thames Valley. Brunel chose the latter as it allowed for better gradients and would also enable the railway to be extended to Oxford, Gloucester and South Wales in later years.

Brunel's survey was presented at a public meeting at the Guildhall in Bristol on 30 July 1833 at which he spoke, later recording that he had 'got through it very tolerably which I consider a great thing'. He added that he 'hated public meetings – it's playing with a tiger, and all you can hope is that you may not get scratched or worse'. A printed account of what the railway committee called a 'numerous and respectable' meeting was subsequently produced and it noted that the meeting had resolved that 'the establishment of a rapid, certain, and cheap mode of conveyance between this city and the Metropolis will be productive of the most important advantages to the Trading and Commercial Interests of Bristol.'[32] The meeting also resolved that a company should be set up to build the new railway, managed by two committees of Directors in both Bristol and London who were tasked with raising the capital to build the railway and obtaining an Act of Parliament. Without a hint of irony, the committee also reported that they had applied to 'several eminent engineers' and had appointed Brunel and Townsend, and that the 'zeal, the diligence, the ability and other valuable qualities manifested by those gentlemen have given them ample reason to congratulate themselves on their choice'. The end result of the initial survey had been to identify the best line 'with respect to distance, elevation, expense, amount of collateral traffic, the probability of junction with other railways etc.' In addition, Brunel had also provided them with a section of the line, and an outline of the number of cuttings, embankments and tunnels required.

The report also described the northern route already recommended by Brunel, highlighting its benefits in terms of connections with Gloucester and Cheltenham via a proposed line from Swindon, and a potential link to Oxford. No London terminus was yet confirmed, the report suggesting that it might be 'at Paddington, or on some part of the Southern bank of the Thames'. Gradients on the new railway would be modest, although its 'comparative levelness' would be achieved by expensive deep cuttings and several tunnels, although the nature of the soil in the proposed line was 'favourable' consisting of gravel, chalk and freestone. Including contingencies and preliminary expenses, Brunel estimated the cost of this grand undertaking as £2,805,330, a sum that would, in the event, prove to be hopelessly optimistic as the great work progressed.[33]

A bound volume now preserved in the collection held at the Brunel Institute titled 'Bristol Railway 1833-1834' provides a unique insight into how Brunel and his team of surveyors arrived at the total budget for the line; the book is packed with costings and calculations for individual sections of line, particular bridges, embankments, stations and tunnels, land prices and potential compensation to be paid to landowners, excavation costs, locomotives and fixed engines. A number of cumulative budgets are noted, the first undated list totalling £2,682,446.16 while the next noted on 5 June 1833 has risen slightly due to increases in the price of excavation and costs for tunnelling rising from £185,000 to £225,000. The last estimate dated 24 June 1833 matches the one produced for both the July 30 meeting and the first prospectus.[34]

Given that the report would be a helpful tool in generating share subscriptions, it included much information about the potential benefits of the railway both to the two cities being linked by the new line, and to individual shareholders. A warning note was struck when it was noted that despite its success, the Liverpool & Manchester Railway had needed to spend 'eleven twentieths' (55%) of its gross receipts on 'causes chiefly referable to the novelty of the undertaking and therefore to inexperience in its execution and arrangements'. Such a situation on the new Bristol railway would, be the report reassuringly noted, be 'highly improbable'. The committee were, the report noted, confident that 'the estimate of cost is liberal' and under a 'deep sense of responsibility' felt that the new railway would be a safe and profitable investment to those who may embark their property in its construction'.

The report listed no fewer than 30 men who formed the Bristol committee of 'deputies', but this cumbersome group had been whittled down to a smaller group of twelve Directors by August 1833. This group included the original promoters, and the wider group provided a good representation of both political views and commercial interests within the city, with links to the Society of Merchant Venturers, the Chamber of Commerce, City Council and the banking, maritime and manufacturing worlds. All twelve Directors provided initial financial support in the company pledging almost £200,000, five investing more than £20,000 each in shares, a considerable sum, a figure equivalent to more than £1.3 million today.[35]

While the Directors came from a wide range of commercial backgrounds within the City of Bristol, a number of investors in the GWR are known to have had interests in the West Indies and the slave trade and in common with many in the business community in Britain during this period, some had owned enslaved people. Others while not directly involved in the slave trade had benefitted financially from the government compensation paid to former slave owners and their descendants after the passing of the Slavery Abolition Act in 1833. This included George Gibbs, described as a 'West India Merchant', who invested £14,400 and who had also supported the Clifton Bridge project. He would go on to invest in the Great Western Steamship Company, the Great Western Cotton Works and the Bristol Dock Company. Gibbs was a partner in Gibbs Bright, a shipping company run jointly with another GWR investor, Robert Bright, who had pledged £25,900; with interests in the West Indies, he would receive compensation for slaves from plantations in Barbados and Jamaica in 1836 and 1837. The Chairman of the railway committee John Cave, with a £12,800 pledge, and Peter Maze, who invested £23,400 in railway shares, are also recorded as having had compensation. It is likely that other investors in the Great Western Railway and other lines built after its completion would have also had capital drawn from similar sources. The total amount of money pledged by the board while representing a large commitment from each Director was still only 11% of the total needed by the company to build the line and much work would be needed to find enough money to build Brunel's railway.[36]

Within weeks of the Bristol meeting Brunel and seven Directors from the city travelled to London to make contact with a provisional committee that had been set up by George Henry Gibbs, a cousin of George Gibbs who was a Bristol Director and a partner in the Gibbs Bright company. On 22 August, the two groups met formally for the first time at Gibbs offices in the City of London. Brunel, anxious to start on the real work of creating his new railway and clearly bored with the formalities, described the London committee as 'rather an old woman's set' who 'needed somebody to give them a little life and sense'.

The person to do this was Charles Alexander Saunders, the new secretary of the London committee, who was introduced to Isambard for the first time a few days later on 27 August when the two committees met again. Born in London in 1796, Saunders would become a firm friend and ally of Brunel and would be a key figure in the development and completion of the railway. There was an instant chemistry between the two men, and after their first meeting, the engineer noted that Saunders was 'an agreeable man'. The Bristol committee had also recruited William Tothill as their secretary, and both men were, as Channon noted, not merely bureaucrats with limited responsibilities but men of means who were able to move in the 'highest commercial and landed circles'. Both had invested £5,000 of their own money in the railway, and would, with Brunel, spend much time speaking with landowners, peers and commercial interests to smooth the passage of the Great Western Railway Bill through Parliament.[37]

The meeting on 27 August was notable for further outcomes; the first was that Brunel was formally confirmed as engineer to the new railway by the London committee, and also that despite the fact that there were still a number of London Directors to appoint, it was agreed that a prospectus for the new project should be issued as soon as possible to begin the process of raising the capital to build the line. When the prospectus was published a few days later, a full complement of twelve London Directors were named, along with bankers and solicitors in both Bristol and London. The prospectus marked the first public notice that the new line would be called the 'Great Western Railway' rather than the 'Bristol and London Railroad'.[38] Brunel biographers have suggested that the new name might well have been his idea, and while there is no firm evidence for this, the name certainly fitted well with his ambitious plans for his new railway. There may well have been more pragmatic and commercial reasons for the choice of Great Western Railway since the promoters of the new line were anxious not to give the capital markets, where the bulk of the investment would come, the impression that Bristol's trade and economic position was in any doubt.

Just over six months after his initial appointment in Bristol, Brunel was now poised to begin the project that would cement his career; after living a somewhat nomadic existence with his parents and staying elsewhere with friends and family in the capital, he now needed a permanent base in London. An office at 53 Parliament Street was rented to provide him with a place where he and a team of draughtsmen could complete the work necessary to submit a bill to Parliament as soon as possible. The office was in the charge of Joseph Bennett, his Chief Clerk, who would work for Brunel for the next 26 years until the latter's death in 1859. A further and necessary comfort was the acquisition of a carriage to enable Isambard to reduce the use of scheduled coaches and hired horses while he travelled up and down the route of the new railway. In typical fashion, Brunel designed his own britzska, which had seats that could be converted into a bed, and ample storage for equipment needed for the survey, and also the cigars he was now smoking almost continuously. This travelling office became a familiar sight to both Brunel's staff and the navvies who were to build the new Great Western Railway, who nicknamed the black carriage the 'Flying Hearse'. Brunel employed a travelling assistant, Farr, who sat on the dickey on the outside of the carriage as it raced across the country and dealt with landlords and stable boys at the various posting houses and inns the engineer stayed at on his travels; it was also his job to refill Brunel's capacious cigar case that held 48 large Havana cigars.[39] It is not known exactly when he finally took possession of the carriage, but an entry in his appointment diary for 12 February 1834 recorded that he had taken an 'early coach' to Reading, returning to London later, at a cost of £1.15s.6d, which would suggest he was still using hired carriages or stagecoaches. The britzska may have been delivered in the late spring of 1834, since no further references to charges for hire of carriages or horses then appear in the diary for the many long journeys the engineer was now making all over the west of England as he began the task of surveying and designing the Great Western.[40]

2

Toil and Trouble

While the new Great Western Railway company was the brainchild of commercial interests in Bristol and London, and its funding would be raised from commercial and private sources, the GWR and other railway companies of the time were 'creatures of Parliament' as their powers were derived from private acts which gave them the right to compulsorily purchase the land needed for the line, compensate landowners and tenants, and raise share capital through limited liability. Parliamentary standing orders provided detailed instructions on what was required before a bill could be applied for; plans and sections of the line needed to be deposited not only with Parliament, but any county or municipal authorities across which the new railway would run, along with notices in local newspapers about the proposed railway. In addition, estimated costs and a list of the landowners and tenants who might be affected by the plans were also required.[1]

Private bills such as those relating to railways were usually considered by Parliament after its Christmas recess and as a result, the documentation for the GWR bill would need to be complete by the end of November to enable members of Parliament to debate the bill early in 1834. With both committees in place, and with the coordination of Saunders and Tothill, further professional assistance was now required from a team of clerks, solicitors, land agents, parliamentary lawyers and engineers to complete this work in quick time. Brunel was formally instructed in September 1833 to carry out a more detailed Parliamentary survey of the line for the railway, to expand on the rapid assessment of the route he had undertaken a few months earlier. The Bristol committee had confirmed on 4 September that it was intended that an application to Parliament should be made in the next session, and that it appeared that from a statement from Brunel 'an expenditure of £500 or £600 at the present time would save both time and enable him to effect that object'.[2] Three days later the London committee also resolved that 'Mr Brunel' should be instructed to start the survey 'without delay';[3] the same day Saunders wrote to the engineer to confirm the resolution of the committee and added that he had been 'directed to beg that you will lose no time' in completing the work.[4]

With no time to lose, Brunel was hard at work straight away. Two days after Saunders letter, his diary read 'Up at 5'and he began the task of traversing the route again. While the basis of the survey was Ordnance Survey maps, there was no substitute for a careful study of the features of the landscape across which the railway was to run, undertaken in all weathers and across all types of terrain from fields to windy hillsides. The 'tools of the trade' were staves, chains and theodolites, all carried miles by the surveyors. Conder[5] noted that having the correct clothing for survey work was important and that two suits, a greatcoat, a waterproof cape and a travelling rug were required to cope with all weathers, a pronouncement confirmed by Thomas Hughes who had already worked with Brunel on the preliminary survey and who reported in a letter to the engineer in 1833 that he had arrived at Wantage 'dripping wet, having stood in the rain two or three hours'.[6]

For the new survey Townsend was instructed to work at the Bristol end of the route with Hughes at the London end, but with so much work to be done in such a short time Brunel was forced to recruit more assistants to help complete it. Aware that some sensitivity might be required in crossing large sections of private property where landowners might not necessarily welcome the idea of a railway, Brunel issued instructions to his staff reminding them to 'ascertain the names of the owners and occupiers of any land to be passed over and as early as possible obtain their sanction to making the survey'. Admitting that this might not be possible until a surveyor was actually on private land, he urged them to make contact with owners as soon as possible. Optimistically he noted that 'I think you will not find any vexatious opposition thrown in your way,' but that surveyors should provide landowners with as much information as they could 'to the manner in which the line may be carried so as to be most advantageous to their property', concluding with the instruction that staff should be particularly careful that 'all persons employed...shall conduct themselves with propriety and civility that no damage be done to the hedges or fences and that they do not unnecessarily encroach upon gardens or enclosed grounds or otherwise annoy the inhabitants.'[7] Surveyors did not always receive a warm welcome. Railways were still very new and many in the countryside regarded steam locomotives and the idea of travelling by train as a dangerous and frightening prospect; crops might be ruined, livestock injured or killed, and fields, woods and pastures divided. When Brunel's staff came 'spying and measuring' it was not just farmers or landowners that they might encounter, but agricultural labourers whose instinct was to drive them off the land they had tended for so long.[8]

While Isambard might not have faced many physical challenges, he had the difficult task of speaking with landowners, many of whom were also opposed to the passage of the railway across their land, not just because they fundamentally disagreed about railways in general, but also with a view to potential financial losses they might incur. It was vital that the issue was addressed with some urgency, and at the end of September, Charles Saunders wrote to Brunel, urging him to ensure that he and staff working on the Parliamentary survey should 'furnish as full a list of the land owners in the line of our railway as possible...with any general information they can furnish as to the opinions & wishes either of them or of the occupiers of property'.[9] What one writer called 'the dogged resistance of the landed proprietary' made life difficult for Brunel, but as L.T.C. Rolt notes, for someone so 'impulsive and forthright' he was able to pacify most and even persuade some to consider investing in the new railway. On 14 September 1833 he visited Mr Wilder of Purley Hall, who he found 'very civil', giving him a prospectus before riding on to a nearby farm to speak to another landowner.

Brunel supervised the team of surveyors, checking plans and sections, estimates and also managed their expenses. Always keen to reduce expenditure, he questioned an expense claim by Hughes who had been working hard, but clearly not hard enough for Brunel, who suggested that if Hughes sent his work to Parliament

Street to be drawn up, it would save time and reduce costs. Hughes did not agree and argued that he would not submit his work 'to be plotted at a cheap rate by another person'. Continuing that he had managed to survey ten miles in a day, although the work 'could not be depended on when done so hurriedly'. Brunel paid the £5 requested by Hughes for the cost of accommodation at various inns three days later.[10]

During the process of completing the survey the long days and nights took their toll and even the indefatigable Brunel was forced to admit to one of his assistants, Hammond, that 'Between ourselves it is harder work than I like. I am rarely much under twenty hours a day at it.' While Brunel was toiling in the field, William Tothill and Charles Saunders were equally as busy organising public meetings in most of the major towns along the route and canvassing the most important commercial interests and manufacturers to both elicit investment and support for the proposed bill. Saunders in particular was 'indefatigable in his efforts to place the company on a firm footing', meeting with most of the members of the House of Commons and a number of peers in an attempt to get as much support as possible for the railway. This tireless work was to no avail; progress in raising capital was slow, and by October less than half of the funds needed to enable the bill to be presented to Parliament had been pledged by investors.[11] Rather than miss the deadline for the next session of the House, it was decided that powers would only be sought for the Bristol to Bath and Reading to London sections of Brunel's railway, with the hope that if this scheme was approved, then further investors would provide the remaining capital to complete the line. Speaking at a public meeting in Bristol later in 1834, William Tothill explained that the decision had been 'a matter of anxious deliberation' and that it had been judged that the most serious consequence of not proceeding was that the impression might have been given that the Directors would be abandoning their original aim of building the whole line in the long term. He added that the board had felt that the approval of the limited bill would be 'readily obtained' and the 'execution of the middle section would be rendered both certain and easy'.[12]

The Board met on 18 October and Brunel was instructed to stop surveying work on the middle section between Bath and Reading, and to 'discharge the assistants' he had employed to help him complete the work up to that point. A revised prospectus was published on 23 October with an additional explanation that noted the reduced scope of the application to Parliament, concluding that this measure would make 'the ultimate completion of the whole line more certain, upon a further application to Parliament in the following year'. The plans and book of reference for this limited scheme were deposited in Parliament on 30 November; the main route of the railway resembled much of the line as eventually completed in 1841, with the exception of its entrance into the capital and the location of the London terminus. Brunel's plan was to enter London on the south side of the Thames with a terminus at Vauxhall. Trains running east out of London would traverse a four-mile viaduct that was to be built with a 6 ft 6 in parapet added to prevent curious passengers looking into houses either side as they travelled through Pimlico, Brompton, Hammersmith and South Acton. The railway then passed under a short tunnel south of Ealing before joining what was the route of the Great Western main line to Reading. Another tunnel was planned near Reading, included to pacify Lord Palmer, a determined opponent of the railway across whose land the route would run. Brunel's original route of the Bristol-Bath stretch also included four crossings of the River Avon, a number reduced later to two. Finally, a branch line to Windsor was included in the Bill but abandoned at an early stage following vociferous opposition from Eton College.

With the Bill deposited, Brunel continued to work at pace and his appointment diary for 1834 provides ample evidence of just how busy he was. Although the survey between Bath and Reading had been abandoned for the present, he still spent much time on the road at either end of the line in the early part of the year. In January he spent days at Acton, Maidenhead and Reading and on Friday 7 February he rode out to Southall and Hanwell before taking a coach to Bath. The next day he met with the Trustees of Roads in the city and on Monday 10 February was 'engaged on the line examining soils' and 'attending at Bath to meet Lord Manvers agent'. He then travelled 'up to town' that evening. The appointment diary was also used to record his expenses, and so notes that his bill at the White Lion Hotel in Bath was 16s. 6d. It seems likely that he spent Sunday 11 February visiting colleagues and friends in Bristol, since also recorded was the expense of hiring a 'Chair to Clifton' at £1, 5s 6d. The diary has many references to Brunel 'working on estimates' as he and his assistants toiled to finalise the cost of the new railway; in a number of other entries the fact that he worked all night was recorded, as were occasions when he travelled all night by coach between London, Bath and Bristol. The work of persuading landowners and aristocracy also continued, with Brunel meeting Lady Berkeley on 5 February and Lord Orkney a week later on the 12th and the following day accompanying Charles Saunders to see Lord Cadogan in London.[13]

The Great Western Railway Bill was one of seven railway schemes introduced into the House of Commons that session. Those opposed to the new railway were a well-organised lobby, illustrated by the circular sent on 8 March 1834 entitled 'Opposition to the Great Western Railway'. It noted that 'the Duke of Buckingham, Countess Berkeley; Earls Jersey, Harrington & Cardigan; Lords Boston, Montague & Stowell; Lady Carr; Mr Sloane Stanley, Col. Gore Langton, Mr R. Palmer, the Provost, Fellows and Masters of Eton College, and the other opponents of this measure earnestly entreat the favour of your attendance at the House of Commons, on Monday the 10th instant, at 12 o'clock, on the motion for the second reading'.[14] The House debated the bill for several hours but despite opposition, the second reading was nevertheless carried by 182 votes to 92. Among those who supported the bill was Daniel O'Connell, the Irish liberal catholic reformer, who argued that the proposed GWR would bring 'great advantages upon Ireland, the great granary and feeding farm of this country'. The debate presaged some of the discussions that were to be had in the committee stage in the coming months; opposing the Bill, Mr R. Palmer MP noted that there was 'much apprehension, he was aware, existing in the minds of honourable members upon the subject of this railroad, and until within the last few days it was not in the knowledge of honourable members generally what was really the Bill before the House'. It was not the case of a railway from London to Bristol he argued, 'but a Bill of very different nature'.[15] Objections from Eton College, to be debated at great length in the committee stage that followed, were also mentioned, although the MP for Berkshire, Charles Russell, described them as 'preposterous', adding:

> No man could take a greater pride than he did in these noble institutions – they constituted a great national characteristic – they went far towards forming the high-minded, and high-spirited character of a great and important branch of the community; and he would not lightly therefore, subject them even to inconvenience. But could it be seriously contended that they were to stand in the way of a great national work?

On 10 April 1834 the bill went to its committee stage, chaired by Henry Granville, Lord Somerset. The stage was then set for a marathon fifty-seven-day battle, one of the longest railway parliamentary committee processes ever seen. As the committee began their work, the company embarked on a further round of lobbying, anticipating strong opposition to their plans from many quarters. There had already been some indication of the coming storm when the number of petitions presented to Parliament for and against the bill had been revealed; the GWR promoters had presented their own petition for the bill on 20 February 1834, one of twenty-three in favour of the railway, but there had been fifty-one against. While there was strong civic support from Bristol and from towns and cities in areas close to the new line such as Bath, Cheltenham Cirencester and Reading, along with places further afield that would eventually benefit directly from the railway such as Neath and Taunton, those ranged against the bill were a different proposition. Almost half of these opponents were individuals, often members of the landed aristocracy, while the rest predictably included businesses that stood to lose from the railway such as canal companies and commissioners of roads.[16]

By the beginning of April 1834 some of the lobbying that had taken place seemed to be working, and it was reported that 'opponents were not so confident of beating us as they were – they say we shall get through the Commons and thrown out in the Lords', a prediction that turned out to be uncannily accurate in the end.[17] The committee stage began, with opponents and supporters of the bill anxious to state their case. No official account of proceedings has survived because three days after the end of proceedings, the old Houses of Parliament at Westminster burned to the ground; extracts from the minutes were subsequently published by the Great Western Railway after the event, but not surprisingly these tend to concentrate on the advantages of the railway and omitted much of the criticism of the Bill. Fortunately, the interminable committee stage was reported widely in newspapers of the time.

Much evidence was presented in the early stages demonstrating the advantages for trade and commerce, and witnesses included the Mayor of Bristol Charles Walker, who told the committee that the GWR would be 'decidedly advantageous to the commerce of Bristol' and that the value of property would be increased. Other merchants and businessmen from the city argued strongly that better communication with London would particularly improve the prospects of the port of Bristol. Thomas Miller, a clerk in the Custom House, reported that the chief imports through the docks were sugar, spirits, tallow, corn, timber, brimstone and wine, and exports were iron, tin, refined sugar, and glass. Without the new railway, he reported, Bristol would 'sink in the scale' and its imports would be materially injured. George Lunell argued that trade between Bristol and Southern Ireland would be benefited by the railway; asked whether the Irish steamer trade and business to South Wales would be increased by 'speedy communication' between London and Bristol he replied that there was 'no doubt of it, that is what we want to increase, the communication'.

Some of the strongest opposition to the Bill came from the Kennet & Avon Canal, which stood to lose much from the opening of the railway. As a result, a number of traders and farmers who could testify about the poor communications between the West Country and London provided by waterways were called by the GWR. William Hier, a sugar refiner from Bristol, and Thomas Stone, a tea dealer, both complained of losses and long delays on the Kennet & Avon which had caused them to move to road waggons, even though it was cheaper by canal. Bath grocer James Shepherd and Thomas Vesey, a tallow chandler and soap maker from Box, both complained about the slowness of the canal, and also the losses through pilferage or robbery they had sustained. Samuel Provis, formerly a freight carrier by canal between London and Bristol, told the committee that the fastest trip between the two cities was around three days, but that frequent seasonal delays, shortages of water in the canal in summer, and winter floods and frosts meant that the journey could take two to three weeks or longer. Richard Mills, a barge master from Reading, told MPs that the longest frost on rivers and canals he had encountered had stopped passage for nine weeks. Farmers, stock breeders and other members of the agricultural trade also had much to say in favour of the new line, even though a complete railway between Bristol and London was not yet a prospect; of the pigs brought from Ireland to Bristol in the 1830s , more than 15,000 were slaughtered there every year and then sent to market in the capital by road wagons, and large numbers of cattle and sheep were driven large distances 'on the hoof' from the West Country to markets in London, a slow, inhumane and expensive process.

After discussions about the public benefits of the Great Western Railway, Brunel took centre stage for much of the committee stage, being the company's principal engineering witness, answering questions on all manner of subjects. These varied from the very technical to the ridiculous, and Brunel displayed remarkable patience throughout, only occasionally making sharp responses to patently absurd requests such as the occasion when he was asked 'Have you selected, in your judgement, the best line of communication?' Brunel retorted 'Of course the line I have selected I consider the best.' It should be remembered that railways were still in their relative infancy and as a writer later noted, parliamentary committees at the time had 'a child's knowledge of railways'.[18]

Much committee time was spent on examining the proposed route of the line, the nature of the gradients, or 'inclines' as they were known, the costs of excavating cuttings and building embankments, and the sizes and mileages of the work involved. The design and cost of almost every bridge and tunnel was examined. Brunel patiently but clearly explained his estimates and route for both sections of line then proposed, demonstrating an encyclopaedic knowledge of his work; given that he had spent almost every waking hour in the past few months working on the proposals, this was not perhaps surprising, but his skill, calmness and familiarity with the detail for someone still relatively young and inexperienced impressed many. Brunel told the committee that his costings were 'a liberal allowance, beyond the prices that have been paid', but also that he had allowed a 25 per cent contingency for any unexpected costs.

To assuage concerns of wealthy landowners, many of whom were strongly opposed to the new line, when asked if in designing the route it had been part of his instruction from the GWR to damage 'ornamental property' as little as possible, Brunel responded 'Most particularly, and I was not only instructed to do so, but endeavoured to do it from the commencement, before I had regular instructions.' Tellingly, however, when asked if this had added to the estimate of the cost he added, 'Decidedly so'. It was said that Brunel's evidence covered more than 500 pages, and the appointment diary kept by Brunel's office recorded that he was giving evidence to Parliament from 16 to 30 April. Following this ordeal, he was then subjected to a further eleven days of cross-examination from 1 to 10 May, and on a final day on 23 May. This examination was undertaken by seven legal counsels in a committee room crowded with landowners and others; the most aggressive of whom seems to have been Serjeant Merryweather, who Adrian Vaughan wrote 'attacked him like a criminal'.

Brunel was asked if he had been in contact with 'experienced engineers' who had been 'practically employed in works of this sort'. Replying positively, Brunel and the railway company were subsequently supported by a number of the most famous engineers of the day, called to give evidence in support of the bill. George Stephenson, asked whether he thought the selection of the route, the size of the estimate, and the 'general conduct of the line' were in his opinion 'judicious in Mr Brunel's hands', told the committee 'I do'. Joseph Locke, then engineer of the Grand Junction Railway, was questioned about the successes of the Liverpool & Manchester Railway and he praised the line, telling the committee that travellers on the line could count on the time of their arrival with certainty, 'generally to a few minutes'. Members of the committee enquired whether there was any 'difficulty or inconvenience attending the travelling by night'. Locke replied that he 'never saw any or heard any' and was able to reassure them that he had travelled by train at night 'often'.

After a few days of Brunel's cross examination, the Chairman called on the promoters of the GWR bill to declare whether they were intending to continue with the proposal to locate its London terminus at Vauxhall. Opposition from landowners and noblemen between Brompton and Chelsea was fierce and the counsel for property owners in Brompton objected strongly to the nuisance the construction of the railway would bring, describing the place as 'the most famous of any place in the neighbourhood of London for the salubrity of its air, and calculated for retired residences'. Another source of opposition from the well-heeled residents of this fashionable part of the capital was the probability that the pigs and other animals which would form part the traffic on the railway would have go through Piccadilly. This, a later report noted unsurprisingly, 'roused the possessors of property in that neighbourhood'. The Commissioners of the Metropolitan Roads were also described as being 'unfriendly' to Brunel's proposal to build a viaduct crossing Piccadilly and had mounted 'strenuous opposition' to the idea. The company agreed to abandon this plan and proposed that the railway would instead terminate at South Kensington near the Hoop and Toy public house, although this plan was further modified in 1835.

While this concession avoided an expensive and difficult stretch of line that would have passed through a fashionable part of the capital, saving the company more than £80,141 of construction costs in the process, it did not seem to assuage the opposition of the noblemen of Brompton and Chelsea who sat in Parliament, and it also further weakened the argument for the railway's already partial scheme.

There were other significant opponents to the railway, particularly in Windsor; in November the previous year, the Provost of Eton College Dr Goodall had made a speech in which he told his audience that he would consider himself 'criminally wanting in duty' if he did not declare that the new railway would be 'ruinous to the school over which he presided'.[19] The boys could easily take themselves well out of the reach of the school and the railway would be dangerous to the morals of pupils he added. Writing later, the historian John Francis added that 'anyone who knew the nature of the Eton boys would know that they could not be kept from the railway.'[20] The authorities of the College remained an implacable opponent of the bill, with Eton being described by one correspondent as the 'fostering mother of almost all the aristocracy of wealth and birth that proud England can boast of'. There is little doubt that the large representation it would have had in Parliament itself was a factor in the eventual defeat of the original

Great Western Bill. A letter published in the *Berkshire Chronicle* also warned residents that any proposed railway would bring all manner of unsavoury people to 'behold the seat of Royalty' from London; the peace of the Borough might be disturbed by 'all the ladies of St Giles and the Seven Dials...and others of the same grade from Old Drury, Field Lane, Billingsgate, Rag Fair etc' it cautioned.[21]

There was further opposition in Windsor prompted by the fact that Brunel's new railway was not to directly pass through or run close to the town; civic pride was further discomfited when it was discovered that the deposited plans showed that Windsor would only be linked to the capital via a branch line from Slough, prompting considerable anger amongst the municipal authorities and local residents. Reporting that the GWR would bypass the Borough, the *Windsor Express* newspaper claimed that the promoters of the new railway could hardly declare that it was a project of national importance if it did not run through Windsor. When a rival scheme was proposed for a direct railway between Windsor and London, there was considerable interest and a petition in support of the London & Windsor Railway Company was presented to Parliament early in 1834, signed by many of the towns leading citizens. William Habberley Price, who had been responsible for a number of earlier schemes for a Bristol-London railway, was commissioned by the Earl of Jersey, a prominent supporter of the proposed Windsor railway and opponent of the GWR, to report on the merits of both railway schemes. While Price argued that he would 'coolly and dispassionately' give his opinions, the predictable conclusion of his report was that the Windsor scheme was 'immeasurably superior to the Great Western line to Reading'. He did however add the caveat that if the Bristol scheme was 'persevered in' then the 'Windsor gentlemen have no other course than to withdraw their Bill this session.'[22]

A further pamphlet published the same month argued that Brunel's railway linking Reading and Windsor would also cost over £300,000 more than the Windsor company proposal, which it argued would be 'cheaper-better-shorter-more level and less opposed'.[23] In the event, Price's final comment would prove to be correct, and the Directors of the London & Windsor withdrew their Bill not long after, largely due to difficulties in raising enough capital to build the line, a position not unknown to their opponents, the Great Western Railway.

A more serious threat came from the London & Southampton Railway who were in the process of promoting a rival scheme to build a railway from Bath to Basingstoke via Newbury, which would then access the capital via their own line to Reading. Brunel's diary records a meeting with the 'Southampton opposition' on 14 April 1834, which may have been intended to discuss some kind of accommodation whereby the Great Western line would join their route at Basingstoke. Already committed to Brunel's original route, the GWR Directors were in no mood to compromise and rebuffed the approach. The London & Southampton instead moved ahead to commission a survey for their own Basing & Bath Railway and returned to oppose the Great Western in the coming months.

The unprecedented length of proceedings for the Bill meant that the House of Commons was obliged to pass additional motions four times to enable the deliberations of the committee to continue. The second reading of the Bill in the House of Lords finally began on 25 July 1834 with Lord Cadogan speaking for the opposition: the Peer highlighted the issue that underpinned much of the opposition to the Great Western proposals, the fact that the Bill was only a partial scheme. He argued that it was 'an incomplete measure' and that there was quite a difference between the 'high sounding' title of the Bill and what it could actually achieve. He and many other critics believed that it was 'a gross deception, a trick and a fraud upon the public, in name, in title, and in substance'. Others suggested that it was neither 'Great' or 'Western' and was 'the head and tail of a concern, 72 miles apart, which would never be joined by a body'.[24]

Despite the assurances of the Great Western Railway Directors there was also still much scepticism as to whether the remaining Bath-Reading section would ever be completed, although Lord Ellenborough told the House that supporting the Bill, he personally could not believe that subscribers, especially those living in Bristol, would have risked such an investment 'unless they felt confident that they would ultimately succeed in the formation of the whole line'. Peers were nevertheless unconvinced, the Earl of Radnor arguing that it was about the wildest scheme he had ever heard of in his life.[25] As a result, in spite the length of Parliamentary deliberations, the Bill was finally defeated by forty-seven votes to thirty. Brunel's appointment diary simply records 'Bill Thrown Out of House of Lords', but a few days later, he did write a brief note reflecting on the summing up of Mr Joy, the Counsel for the Parliamentary Bar. His observations were, Brunel thought, 'rather unparliamentary' and noted that Joy had stated that the railway was '"Not great – except in conception...Not Western, Not a Railway, but two"'.[26]

For some, the failure of the Great Western Railway Bill was a cause for celebration; soon after the event, the Marquis of Chandos presided over a large public meeting at Salt Hill in Slough to mark the defeat; since Salt Hill was a busy location on the Great West Road, where sixty coaches might pass each day, and where several thousand horses for carriages and chaises were stabled in the area, the rejoicing of those whose livelihoods depended on equine transport might be forgiven. The press also took the opportunity to pass comment; one commentator known as 'Cort' noted 'The Great Western, though probably it may reach as far as Bath from Bristol, after having, like a mole, explored its way through tunnels long and deep, the shareholders who travel on it will be so heartily sick, what with foul air, smoke and sulphur, that the very mention of a railway will be worse than ipecacuanha.'[27]

Brunel, Saunders and the Directors were undeterred, and efforts continued to be made to raise enough capital for the complete scheme. In September 1834, Saunders wrote to existing shareholders to reassure them that despite the loss in Parliament, the Directors were determined to succeed. The evidence already presented to Parliament had established not only the importance of the new line he argued, but also the accuracy of both potential revenues to be generated and the 'sufficiency of the estimates of cost', the latter claim being in retrospect wildly optimistic, given the fact that Brunel's line would eventually cost more than two and a half times the original estimate! There was also a tacit admission that the earlier strategy of only applying for permission to build the Bristol-Bath and Reading-London sections of the route had been a mistake; this proposal had been the only factor with a 'reasonable ground of objection' and the promoters 'could not disguise from themselves that it was but a *partial* benefit compared with the whole line'.

The Directors now felt it was their duty to complete an 'adequate subscription' of shares to enable them to submit a bill for the entire line in the next session of Parliament. Concluding his letter, Saunders offered existing subscribers the opportunity to buy further shares or dispose of their current allocation. A new prospectus was issued in November, with an estimated cost of £2.5 million, a sum required to construct the whole line, including depots, stations and locomotives. The London terminus was now intended to be at Euston, near the 'New Road in the Parish of St Pancras', to be accessed via a junction with the London & Birmingham Railway at Wormwood Scrubs.

Determined not to fail again, Brunel and the Directors now launched a charm offensive, holding public meetings to canvass further support at locations in all towns of importance in the West Country. Petitions to Parliament were also circulated in other more distant areas, to gather as many signatures as possible in favour of the new Bill. A flavour of one of these events can be gained from an account of the public meeting held at the Merchants Hall in Bristol on 15 October 1834.[28] Saunders and Brunel would have been in little doubt as to the importance of maintaining and growing support in Bristol, the place where, after all, the idea for the railway had originated. The Chair, Charles Payne, set the tone by stating in his introduction that if the people of Bristol 'did not exert themselves, they will probably be distanced in the race of commercial prosperity'. After various resolutions of support from civic organisations such as the Dock Company, the Society of Merchant Venturers and the Corporation, the somewhat chequered history of the company was recounted by William Tothill, Secretary of the Bristol committee. Lessons had been learned he argued, although the loss of the first bill was to be 'regretted on account of the loss of time, and of a large expenditure, which has not led to immediate success'. Much experience had been gained and not only had the mass of evidence presented conclusively established the case for the railway, the means of 'conciliating opposition' and of obtaining the support of important interests meant that the success of a new bill was, Tothill claimed, 'certain and will be speedy'.

There was more evidence that with a Bill for the whole route in prospect many landowners had now softened their attitude to the railway, it was noted that the Duke of Buckingham who had opposed the railway, had now 'become their friend' and that two other noblemen, the Duke of Cleveland and Lord Manvers both of whom owned property in Bath, were now in favour. Captain Lye from Bath told the meeting that he had been authorised by Lord Manvers 'to state that in the event of the whole line being applied for, he would not offer any opposition'.

The Chairman then introduced Brunel, telling them the engineer was 'labouring under indisposition' but would nevertheless provide more information relating to the construction of the railway. Brunel replied that 'he would endeavour briefly, but he was afraid imperfectly', to bring before the meeting the principal features of the line. The facts might be 'dry and uninteresting' he added, but he hoped 'his manner would not render them more so'. Telling the assembled meeting that in the process of building a new railway 'it was obviously impossible to please everyone,' he described the reasons for his choice of route. His preference was he said driven 'only by public reasons' with his line offering the 'quickest transit', and including the greatest number of large towns. In recommending the northern route, he noted that while a railway built on or near the course of the Kennet and Avon Canal could have given the best line, this was not the case and a route north of the canal would be better for engineering, and brought them much closer to the towns of Oxford, Gloucester and Cirencester and in addition, it was planned to build a further branch from Chippenham linking Bradford and Trowbridge. Brunel was also quick to mention that 'Mr Stephenson, the father of Railway Engineers, stated in evidence that no better line of country could possibly exist.' Carefully not mentioning the fierce gradients that would have to be overcome at the western end of the line, he reminded the meeting that on the section of his railway from London to Swindon, a distance of almost 70 miles, there was only an ascent of 6 or 7 feet in a mile, and from London to Reading it was almost a dead level. It was, he concluded, 'evident from this that the railway could therefore be formed at a cheaper rate and offering greater advantages in the shape of rapid and cheap transit of goods and passengers on this line than any other in the country'.

Brunel also had something to say about the opposition he and others from the GWR had met with from the landowners on the line and the progress that had been made in conciliating and reducing opposition to the railway. He was quick to point out that the Great Western appeared to have more support than was the case on some of the other large lines then being promoted. This did not mean to say that there was no opposition he added, since 'some persons disliked having railroads through their estates,' concluding that the Directors wished, however, to 'avoid as much as possible interference with private property'.

Whatever kind of 'indisposition' or illness seemed to be troubling Brunel that day, it does not seem to have had reduced the impact of his contribution to the public meeting. After railing against a campaign of disinformation that he claimed had been waged against the railway through misstatements which had been printed in many of the newspapers, and 'otherwise industriously circulated through the country' he ended his speech to cheers from the Bristol meeting when he told them that the opposition had been 'overcome' in many places, and it only needed a little energy from the people of Bristol and the shareholders in general 'to ensure the successful completion of this great national undertaking'.

The very positive mood the meeting had enjoyed up to this point was then somewhat cooled by the intervention of a delegation from Windsor that included its two Members of Parliament, intent on pressing their case for a more direct connection with London via Brunel's main line. This interruption was dealt with by comments from shareholders firmly in favour of the route as proposed, and Charles Saunders, who reminded the meeting that they had just heard from Brunel that George Stephenson had considered the best line was the original one chosen by the Directors, and that the Great Western was more than happy to provide Windsor with its own branch line. This distraction dealt with, Saunders brought proceedings to a close; surely, he argued, 'there could be found in this city gentlemen to take ten shares each, and then the work was done.' Finally, he called upon the meeting and the people of Bristol, 'cordially to unite, and vigorously join hearts and hands, in and for the completion of the Great Western Railway'.

If the account of this meeting was typical of others held across the West Country, then it is not surprising that both Brunel and Saunders found the whole process so demanding; and they were still engaged in the task of speaking with individual landowners and opponents of the Bill at the same time. With more support being shown by organisations such as the Society of Merchant Venturers, more manufacturers, traders and small businessmen were persuaded to invest. This included a further group now making use of compensation granted to them following the passing of the Abolition of Slavery Act in 1833, such as Philip Miles, the MP for Somerset. Personal contacts were a part of the process of generating the necessary capital and the Directors in Bristol and London called on their network of acquaintances across the length and breadth of England and Wales, although a large proportion of the original subscribers were ultimately drawn from areas closest to the route of the railway – London, Berkshire, Buckinghamshire, Oxfordshire, Gloucestershire, Wiltshire, Somerset and Bristol.[29]

By February1835 Charles Saunders was able to confirm that the 10,000 additional shares required by Parliament had been raised, allowing a petition for a second bill for the construction of the whole line to be submitted. A handbill issued by the Great Western proclaimed that the railway, 'objected to last year in consequence of its incompleteness, is now brought before Parliament in a perfect state'. The document also reported that 'the feeling of the landed Proprietors whose estates would be intersected by the Railway is also decidedly in its favour.'[30] This naturally gave the Bill a greater chance of success in Parliament, although the increasing numbers of railway schemes being promoted and reports of potential income to be generated from these must also have been a factor.

After the prolonged discussions that had characterised the first bill's passage through the House of Commons committee, Brunel, Saunders and the Directors might have hoped that the new proposal would receive an easier ride. The chairman, Charles Russell, the Member of Parliament for Reading, began by telling the committee that the deliberations undertaken the year before had proved the 'general utility' of a Bristol to London railway, so it would not be necessary to revisit many of the basic arguments that had been debated for so long previously. This was a major victory, weakening the position of many of the opponents who had already stated their case in the committee stage of the first bill, such as the Kennet and Avon Canal Company.[31] A petition had, however, been made in 1835 by coach masters and owners of goods wagons. In their submission to Parliament, they argued that the new railway would restrict competition and create a monopoly; this would mean that 'the present trade of your Petitioners will be much lessened, if not wholly destroyed,' they wrote, a prediction that would unfortunately become a reality in a few short years.[32] There were around twenty coaches running daily between the two cities at the time the GWR Bill was being debated, the fastest coaches making the journey from Bristol to London in around thirteen hours, although many took much longer. Passengers were charged around 40 shillings for an 'inside' seat, and half that for a chilly 'outside berth'. The journey could be long and uncomfortable, especially in the winter, and not always safe; the *Railway Times* reported regularly in the 1830s and 1840s on 'Coach Accidents' recounting serious incidents and near misses on the road with some glee. The transportation of goods by road was a far slower process; wagons weighing around a ton and a half lumbered along the Great West Road, and it could take almost a day just to travel from London to Reading. Clearly it was not only coach or wagon owners who stood to lose, but also stables, fodder merchants and coaching inns and hotels along the route, so their opposition was inevitable.

Although Russell's intervention had removed some obstacles and prevented old arguments being rehearsed yet again, committee proceedings still extended to forty days. The London and Southampton continued to be a fierce rival as it had been during proceedings for the first Bill, and Brunel told the committee that the Great Western route was 'evidently superior' to the Basing & Bath scheme proposed by the London & Southampton Railway that continued to be promoted as a better alternative to his railway and moreover, it would not open 'any communication between the South of England and the Midland Counties', he concluded. In the gap between the first and second GWR bills, the Basing & Bath promoters had organised a number of public meetings in support of the proposed line. The Duke Street appointment diaries record that Brunel attended a meeting held in Bath on 12 September, which did not go altogether to plan from the organisers' point of view.

For a start, there was a good deal of confusion about whether the rival scheme would actually reach Bristol, which had for many been part of its attraction. When it was announced that, initially at least, the railway would only run from Basingstoke to Bath, and that it would cost one million pounds to build, there was pandemonium in the meeting. Charles Saunders, also present to represent the GWR, was quick to point out that this admission undermined the whole scheme as promoted. Brunel then stood up and proceeded to demolish the arguments put forward in favour of the Basing & Bath route, particularly emphasising the fact that it would follow the line of the Kennet & Avon canal, entailing severe gradients and a tunnel at its summit, whereas his northern route would win both on 'superiority of levels' but also because it would run much closer to potential lucrative sources of traffic. The meeting concluded with a resolution passed to loud cheers in favour of the Great Western Railway Directors and their 'exertions' in linking Bristol, Bath and the capital, not quite the outcome the London & Southampton Railway had envisaged.[33]

Further meetings were held by the Basing & Bath promoters, including 'a numerous and highly respectable' gathering in Frome on 17 December. Brunel was absent on this occasion, attending a meeting of the Thames Tunnel committee in London, but supporters of the rival railway were still in a bullish mood, one telling the meeting that the real issue was 'which was the real Western line to the real West of England' and that the Great Western had 'no more right to this title than any person present'; not surprisingly, there was little mention of Bristol and more emphasis on local benefits to the line.[34]

The London & Southampton opposition continued to rumble on, and eventually the Great Western felt aggrieved enough to publish a short pamphlet as committee proceedings began, rebutting many of what they called 'unfounded misrepresentations', directly naming the London & Southampton as the 'real opposition' which they had to contend with and arguing that their route was 'utterly inferior to that of the Great Western Railway'. It was also reported that the Basing & Bath route, while shorter than the GWR line, would involve the excavation of 24 million tons of spoil and cost £50,000 per mile to build. The pamphlet also repeated Brunel's claim at the Bath meeting the previous year that the gradients on the Basing & Bath were also 'inferior'.[35]

As chief witness for the Great Western Railway in the second committee stage, Brunel was required to repeat many of these arguments and answer more questions regarding the route of the new line. Brunel once again faced the formidable Serjeant Merewether and spent eleven of the forty days during which the committee considered evidence being cross-examined. He reconfirmed that the northern route via Chippenham and Swindon was 'superior on every account' rather than the other proposal that would have seen the railway run south of the Marlborough Downs. Detailed questions were asked about the volume of earthworks required for Brunel's railway and he was able to confirm that there would be a comparatively small number of cuttings which would involve the removal of nearly ten million tons of soil and rock, a third less than the Basing & Bath line, a consequence of the easier terrain his railway would cross and his talents as a civil engineer. The Victorian writer George Sekon reported that a person who was present in the committee stages of the Bill said that Brunel's 'knowledge of the country surveyed by him was marvellously great' and that 'he was rapid in thought, clear in language, and never said too much, or lost his presence of mind. I do not remember having enjoyed so great an intellectual treat as that of listening to Brunel's examination.'[36]

Brunel's opponents were forced to resort to other avenues of attack, in particular his plans to build a tunnel through Box Hill. The committee heard evidence from William Reed, described initially as an 'earth mover' but who, it was later revealed, was actually the Secretary of the Basing & Bath Railway Company. He argued that tunnels were 'always inconvenient' and that

> No person would be shut out from the daylight with a consciousness that he has a superincumbent weight of earth sufficient to crush him...the inevitable and certain, if not the necessary consequence of such a tunnel would occasionally be the wholesale destruction of human life, and that no care, no foresight...could prevent it.

The tunnel was 'monstrous, extraordinary, dangerous and impracticable' in the view of other critics and for many, the idea of travelling through a tunnel of a mile and three quarters was unthinkable, the noise of two trains passing each other in the tunnel would 'shake the nerves of this assembly' one witness claimed, adding that no one would do the journey twice.[37]

The tunnel would also be built on a gradient of 1 in 100 and it was this novelty that brought Brunel into debate with Doctor Dionysius Lardner, an Irish scientific writer who had edited the *Cabinet Cyclopaedia* in 1829, but whose unconventional pseudo-scientific ideas had made him a figure of ridicule to many. Described as 'ignorant and impudent' after speaking at a meeting of the British Association, Lardner was not above criticising many of the most important figures in the scientific and engineering community.[38] He presented detailed calculations to the committee purportedly showing that a train running without brakes down the 1 in 100 gradient would quickly gather speed, reaching 120mph as it burst out from the west portal of the tunnel. Brunel, with some barely disguised irony, pointed out that Lardner had neglected to account for wind resistance or friction, and that the figure he might have come up with would have been 56mph rather than 120mph.

Brunel had told the committee on 25 March that he had considered using a stationary steam engine to help haul trains up the incline rather than use locomotives, but that 'I cannot say which I mean to adopt.' There was a long discussion in the committee about what might happen if the ropes broke, but five days later Brunel was able to call on the support of George Stephenson and Henry Palmer, who had been asked by the Directors to report on the proposed incline through the tunnel. An affirmation of both Brunel's plans for the tunnel and his route more generally was published by the railway and presumably shared with the committee.[39] It stated that a shorter, steeper incline was the one recommended, since it would be suitable for either stationary or locomotive engines. Having examined the route of the whole line they considered it 'judiciously selected' and 'undoubtedly superior to those of the London-Southampton or the Basing and Bath'. In a separate letter to Directors sent at the same time, Stephenson and Palmer told Directors that in suggesting a shorter and steeper incline through the hill at Box, 'Mr Brunel had our concurrence.'[40]

Throughout the proceedings and to reassure the wealthy landowners and peers who had objected so strongly to the line previously, the engineer again told the committee that he had tried wherever possible to avoid damaging land or properties of any importance. In a rather different exchange on 24 March, he was asked about the demolition of properties in Bath. He described the homes as 'wretched…inhabited by lodgers of the worst description'. Brunel's subsequent responses provide a telling picture of his Georgian attitude to social class; asked whether it was the poverty of the people he objected to, he replied, 'No, the class of people', agreeing with the counsel that the unfortunate tenants of these houses were 'not merely guilty of poverty but something worse' and arguing that his railway would be a 'great moral improvement' in the city.[41]

The committee finally completed its deliberations at the end of August 1835. It approved the Bill subject to some amendments to protect the scholars of Eton College; the GWR agreed that no station would be built within three miles of the school, and that a sturdy fence should be built along the line close to it. With all objections finally overcome, the Great Western Railway Bill was returned to the House of Lords for approval and passed into law on 31 August. The birth of Brunel's new railway had not been cheap; at the first half-yearly meeting of the company after the passing of the Act in October 1835 shareholders were informed that out of almost £89,000 spent by the company so far, £18,168 had been used to fund Brunel's survey of the route, land valuation and expenses for professional witnesses in parliament, and a staggering £38,771 had been spent on 'solicitor's bills and other legal fees', the Directors concluding that it was considered by no means 'disproportionate to the object attained'.

3

Building the Line

Following the passing of the Great Western Railway Act in August 1835, one contemporary writer noted: 'No efforts were spared by the Directors and officials of the new undertaking to mature their project in the shortest possible time.'[1] Within a few weeks Brunel had written to his assistant George Frere telling him that he had instructed Townsend to 'obtain immediately permissions to cut sufficiently through the small but thick underwood in Brislington and that neighbourhood' (the Avon Valley east of Bristol), so that the route of the railway could be set out properly.[2] It was clear that the hiatus between the two GWR bills had enabled Brunel to further develop his vision of 'the finest work in England'. For the foreseeable future, while still working on other projects such as the Clifton Bridge, harbour schemes and his first two steamships, the primary focus of Brunel's attention would be the completion of his new Bristol to London line and the creation of a network of other routes that would eventually make up an integrated railway system covering the West Country, the Midlands and Wales.

It soon became apparent that Brunel's new GWR was to be no carbon copy of other existing or planned lines, and would instead be a completely new design, with innovatory architecture, civil engineering, motive power, track formation, and most importantly a radically different track gauge of 7 ft 0 ¼ inch, the 'broad gauge'. Describing the Bristol to London railway after completion, Francis Whishaw noted that in creating his new line Brunel had not been 'satisfied with the beaten track pursued by those who had gone before him, determined on carrying out this important work on entirely new principles'.[3]

Brunel, while aware of other railway developments of the time, was characteristically determined to design and build a line of his own conception and, impatient at the way in which railways were then developing in an evolutionary manner, was instead anxious to improve the design of his line in one revolutionary leap. The railway would be a high-speed route with the best levels, which, apart from two steeper stretches near Dauntsey and at Box, had a ruling gradient of 1 in 660.[4] Working with the form of the landscape, excessive earthworks such as cuttings and embankments were avoided wherever possible on its sweeping curves. Generously engineered structures were an expression of Brunel's vision for a railway on a grand scale. Writing in his 1838 Guidebook of the then unfinished GWR, E.T. Clark noted that 'the unusually favourable gradients, the absence of objectionable curves…it was proposed at a very early period of the undertaking to travel at a higher speed and with greater regard to steadiness of motion than had been obtained upon other railways.'[5]

Much of this grandeur is a result of the larger-scale civil engineering required to accommodate Brunel's broad gauge, perhaps the most radical innovation he proposed for his new railway. The history and fortunes of the Great Western Railway and others eventually associated with it would be inextricably linked with the consequences of increasing the width or gauge of the rails on its lines from 4 ft 8 ½ in to 7 ft ¼ in and Brunel's bold experiment would not only cause him much anxiety and trouble during his career but would also cost the company much financially and in reputation in the coming years.

Giving evidence much later to the Gauge Commission in 1845, Brunel argued that the decision to adopt this novel idea grew upon him 'gradually', but in public at least, he had said little if anything about this revolutionary idea. As a result GWR Directors did not hear of this important development until a month after the passing of the Act of Parliament on 15 September 1835, when he wrote to them recommending that with regard to the track gauge of the new line they should approve 'a deviation from the dimensions adopted in the railways hitherto constructed'.[6] Rolt records that the original 1834 Great Western Bill did have a clause limiting the track gauge to 4 ft 8 ½ in, but when the 1835 Act was subsequently debated in Parliament following discussions with Lord Shaftesbury, Chairman of Committees at the House of Lords, this restriction was omitted at Brunel's request using the precedent of the London & Southampton Railway Bill, authorised in July 1834, suggesting that the engineer had been thinking about this bold new idea for some months.

Until 1835, the question of track gauge had been of little consequence to those promoting, building or investing in railway schemes; what became known as the 'standard gauge' (4 ft 8 ½ in) had been used on most of the new railways constructed up to that point, and had originated largely as a result of precedent based on the fact that most of the railways and tramroads built to serve mines, tramroads and ports in the North East of England had a track gauge of around this, thought to be the width of a cart horse and the wagon it hauled. These developments in what became known as the 'Cradle of Railways' were cemented by the work of George Stephenson, who, after his appointment as engineer of the Stockton & Darlington railway in 1821, chose a gauge of 4 feet 8 in, based on his experience of working in the Northumberland coalfield; Stephenson had designed and built his first two locomotives in 1814 at the Killingworth Colliery both running on a line with a similar gauge and so it was no surprise that he advocated a standardised approach, which became truly standard when it was adopted on the Liverpool & Manchester Railway in 1830, and subsequently on the Grand Junction and London & Birmingham railways.

The organic and evolutionary way in which the standard gauge had developed must have been a factor in Brunel's decision to recommend a wider gauge, but there is no mention of this in his report to Directors, which was instead full of references to friction, resistance and wheel sizes, noting that 'by simply widening the track gauge so that the body of the carriage might be kept entirely within the wheels' the centre of gravity of rolling stock would be 'considerably reduced' and at the same time the diameter of the wheels would be unlimited'.[7] He added that 'I should propose 6 ft 10 inches to 7 ft as the width of rails which would I think admit of sufficient width of carriages for all purposes,' concluding that the motion of carriages would be 'more steady' and wear and tear reduced. There was no mention of other potential advantages, such as larger and more powerful locomotives or bigger carriages and wagons. Brunel concluded by outlining what he called 'objections' to his proposal, namely the increased costs of earthworks, bridges and tunnels which, he argued would 'not be so great as first sight appeared', reporting that slopes of cuttings and embankments would remain the same and that an allowance for the increased dimensions of bridges and tunnels had been 'provided for' in the estimates. This was by no means the end of the broad gauge debate and Brunel would be forced to make a more robust defence of his innovation when the first section of the railway opened three years later.

The most significant problem highlighted by Brunel in his report and what he described as 'the only real obstacle' to the adoption of the broad gauge was the 'inconvenience of effecting a junction with the London & Birmingham Railway'. To call this issue an inconvenience was an understatement. It had been proposed in the 1835 Act of Parliament that the Great Western would share a station at Euston with the London & Birmingham, and the idea of having one large central station for both companies was a far-sighted one providing 'accommodation for through traffic without change of vehicle'. Negotiations between the companies began in 1834 and dragged on. A surviving draft agreement proposed that the London & Birmingham should provide land 'to be sold to the Great Western Company at a fair valuation for the purposes of their depot and station house at the present or any future terminus of the line' and annual rents for access to the station were also to be determined.[8]

The adoption of the broad gauge was undoubtedly a further cause of disharmony between the two companies; in his September 1835 report Brunel had suggested that an additional rail could be laid to enable trains of both companies to share the station, and that he 'did not foresee any great difficulty in doing this, although adding that 'undoubtedly the London & Birmingham Company may object to this.' This rather glib statement was somewhat hopeful since it seems unlikely that Robert Stephenson or the Directors of the London & Birmingham would ever agree to such an arrangement.

Differences and misunderstandings with the Directors and officers of the London & Birmingham continued into 1835 and the arrangement eventually collapsed acrimoniously, as in the event the London & Birmingham refused to offer the Great Western anything more than a five-year lease on land and buildings at Euston. Without the guarantee of a long-term lease or the purchase of land for station buildings, GWR Directors were unsurprisingly reluctant to risk capital on building a junction and new branch line to Euston. Following the protracted discussions, a further final deputation from the GWR met with Directors of the London & Birmingham in November 1835, but no agreement could be reached and at the first half-yearly general meeting of the company in February 1836 it was announced that the proposed arrangement had been broken off and that the Great Western would instead build its own London terminus at Paddington in West London. At the following shareholders meeting in August, it was then announced that they had obtained 'the general consent of owners and occupiers to an extension line to Paddington'. The new railway running from Acton to Paddington was around four miles long and a further bill would be deposited in Parliament for these arrangements.

The report of the second half-yearly meeting of shareholders held on 25 August 1836 also revealed almost in passing the significant news of Brunel's new track gauge. After some glowing reports of progress on the construction of the railway, and an announcement that an opening to Maidenhead might be possible in a year, it was noted that in the course of setting out the route the engineer had been able to make further reductions in the gradients to be encountered. As a consequence of these favourable gradients, the Directors were, the report stated, confident that the cost of locomotive power on the new line could be reduced, meaning that a 'proportionate increase of profits will be received by the proprietors' and train speeds would be increased. On the last paragraph of the report's second page it was then revealed that 'under these peculiar circumstances' and to 'remedy several serious inconveniences' experienced on existing railways 'an increased width of rails has been recommended by your engineer, and after mature consideration has been determined on by the Directors.' There was no mention of what the new gauge might be, or whether there might be any serious financial consequences as a result, although these questions and others would be raised by shareholders in future meetings.

An unknown correspondent for the original *Great Western Magazine* published in 1864 noted that there was 'nothing more thoroughly uninteresting to the general public...than the details of the construction of a railway'[9] and that as soon as the railway was built and opened, the process by which it became a railway was forgotten by most who travelled on it, a phenomenon that persists today with major engineering projects such as the Channel Tunnel and the Elizabeth line. Although it is probably true that passengers ultimately had little interest in the obstacles encountered in constructing the GWR, there is, even today, still much to marvel about in the story of the construction of Brunel's Great Western.

With parliamentary approval a new organisation was required to actually build the Great Western Railway. Although Brunel had employed a number of assistants while working on proposals for railways and other projects such as the Clifton Bridge, it was not until late 1835 when his success enabled him to move from offices in Parliament Street to a much larger house at 18 Duke Street Westminster that his staff increased significantly.[10] Brunel's first appointment was Joseph Bennett, recruited as Chief Clerk, a post he would hold from 1836 until the engineer's death in 1859. Bennett ran Brunel's office, managed his commercial finances and dealt with the copious correspondence to and from Brunel and his staff. Little is known of this long-suffering employee, but his legacy survives in the thousands of letters and documents still surviving in The National Archives and Brunel Institute collections.

What we might now call the railway industry was still in its infancy in 1835, and Brunel and others had little previous experience to call on as he embarked on what was then the largest railway construction project ever contemplated. As we have already seen, Brunel was not always anxious to follow precedent, but clearly the experience gained by George and Robert Stephenson who were then engaged in the most significant railway works up to that point could not be ignored; Brunel knew and respected both men and while they remained professional rivals throughout their lives, Robert Stephenson in particular became a close personal friend and remained so until 1859, when both men passed away within months of each other.

It had been hoped that some of those already working for the Stephensons on railway construction might be tempted to move south and work on the new GWR, but with work proceeding apace on the Grand Junction and London and Birmingham railways and other lines such as the London & Southampton also in progress, this proved not to be the case and Brunel was forced to find and train his own team of assistants and resident engineers capable of matching his high standards; he was able initially to call on engineers who had worked with his father on the Thames Tunnel project such as Richard Beamish, William Gravatt, Michael Lane, Thomas Page and Charles Richardson, while others such as George Clerk, G.E. Frere and John Hammond were appointed soon after work started on the GWR.[11] There was no doubt that Brunel was a hard taskmaster and expected his staff to work as hard as he did and follow, not question, his instructions; his Private Letter Books contain many examples of reprimands and rebukes to those he thought had failed him and while he was prepared to sack errant assistants this was usually seen as a last resort, and those staff demonstrating indolence and laziness usually received a final chance to redeem themselves.

In 1836 Harrison, a sub-assistant employed on the viaduct at Hanwell, was dismissed by Brunel for 'want of industry' but was given the chance to work 'on trial' at the western end of the line for a month with the promise that if he demonstrated 'greater exertions' he might be reinstated.[12] More severe was the much-quoted letter to a recalcitrant assistant named Fripp, whom Brunel described as a 'cursed, lazy, inattentive, apathetic vagabond' whose 'untidy habits' and neglect had 'wasted more of my time than your whole life is worth'.[13] The pressure to deliver was felt keenly by Brunel's assistants as work on the railway progressed; in 1837 Hammond wrote a number of letters to Archibald, another assistant, complaining of delays in completing measurements needed for contracts and being slow to respond to requests for information. 'By all that's good you must get on faster,' he added, concluding 'for the love of fame and our great masters push on the work.'[14].

Brunel also relied on the recruitment of pupils who wished to become civil engineers; in theory these unpaid employees would receive at least some training or supervision from the engineer but in practice they largely learned on the job, being sent to work with Resident Engineers and other members of Brunel's staff. The rare survival of the journal of Charles Richardson (1814-1896) for the years 1835 to 1838 provides a unique insight into the initial stages of railway construction on the GWR and the life of those working on it. Richardson had been born in Cheshire but had moved with his mother to Bristol as a child; leaving university in Edinburgh to take up a new career as a civil engineer he joined Brunel's office at the age of twenty and would work on the Great Western, Bristol & Exeter, and Cheltenham & Great Western Union railways, as well as assisting at times on the Thames Tunnel and Clifton Bridge projects. Before the passing of the GWR Bill in August 1835 the young man spent time on a number of trips with Brunel, known as 'Mr B' in the diary, soon discovering the long hours worked by his boss. On 8 August, he travelled with Brunel to Shapley Heath leaving at 4am and spending all day 'examining cuttings' on the line. Having spent his youth in Bristol and the surrounding area Brunel made use of his local knowledge and Richardson was sent there in October 1835, spending almost a year there helping with the setting out of the route at the western end of the line. Brunel seems to have liked Richardson, who generally worked hard, although even he was the recipient of a strongly worded letter from the engineer when his love of cricket seemed to be taking precedence over his professional responsibilities.[15]

Samuel Smiles description of Brunel as the 'Napoleon of engineers' might well be a little overblown, but it was close to the truth. Brunel was firmly in charge of everything and in the matter of engineering he took full responsibility for the work and would not contemplate its delegation to any subordinate. Brunel later admitted when writing about the SS *Great Eastern* project that he could not 'act under any supervision, or form part of a system which recognises any other advisor other than myself'. As Angus Buchanan noted, it was only the physical impossibility of doing everything himself that meant he had to reluctantly trust others to be his 'eyes and hands' when he was not there.[16] The obsessive need for control and the inevitable long and unsociable hours he was required to spend working would take their toll on his health and also meant that he spent little time at home, particularly during the construction of the Great Western Railway. Surviving appointment diaries chronicled many occasions when Brunel worked long days and also seven-day weeks: an entry for Sunday 10 January 1836 records that 'IKB working all day – various letters to landowners such as Lord Jersey, Duke of Buckingham, Babbage, Mr Osborne on B & E. Business assistants all working too'. Brunel's staff were not always happy about these long hours; a member of staff had annotated the Duke Street appointment diary for Good Friday 1843, adding a handwritten comment to the entry which said it was a 'Holiday at all public offices' adding 'Except Mr Brunel's'.[17] If he was not engaged in meetings or visiting works along the route of the railway he was often travelling by carriage to and from Duke Street. On 7 April 1836 his diary noted that he was 'engaged at Bristol and the line, returned to town at night. Met Davenport at Reading by appointment'.[18]

Mention of 'B & E' (Bristol & Exeter Railway) in the January 10 diary entry also highlighted a further pressure on Brunel, then facing the biggest challenge of his career; not content with his appointment as engineer to the Great Western Railway he had also taken on numerous other railway projects which would eventually form part of his railway empire. The GWR Chairman Benjamin Shaw had recorded in his report to shareholders in August 1836 that bills for railways from Bristol to Exeter, Swindon to Cheltenham and Cardiff to Merthyr had already been passed by parliament and that work on these would be undertaken with 'the utmost vigour and dispatch'. The possibilities of a branch to Oxford and a further line from there to Worcester were also noted. Shaw concluded that it was 'almost superfluous to remark that the Great Western Railway are materially interested in the completion of those undertakings'.[19] Brunel was to become the engineer of all these railways, along with a branch line to Newbury and the poor old Bristol & Gloucester Railway, which he claimed in his diary that he had 'forgotten', a project that nevertheless had a capital of £450,000. It was hardly surprising therefore that Brunel was so seldom at home since he was not only travelling along the route of his Great Western line to Bristol, Bath and Reading, but was also to be found at places such as Exeter, Taunton, Gloucester, Cheltenham and Cardiff.

Brunel had mentioned the Merthyr to Cardiff Railway in the list of projects he was committed to in his Boxing Day 1835 diary entry. He had come to the attention of wealthy ironmasters in South Wales though his work promoting the GWR, and connections he made as he sought iron for his new Clifton Bridge and other projects. In October 1834 he was asked to survey the route of a line from Merthyr to Cardiff to link ironworks to the docks in Cardiff. In the midst of his work on the GWR, Brunel took time to complete the survey, spending gruelling days in the field and sleeping in his carriage at night, once spending two weeks without going to bed.

After completing the survey, there was little progress initially but the scheme was revived the following year and in 1835 the company, now called the Taff Vale Railway, was incorporated.[20] The Taff Vale Act was passed in 1836 with Brunel estimating the cost at £286,031,018.[21] With the Bill passed, Brunel was heavily committed to his work on the GWR and other lines and was in no position to fully supervise the construction of the railway, so the TVR appointed the experienced George Bush as Principal Engineer. Brunel had noted in his diary that he 'did not care much about' the project but nevertheless remained anxious to be 'engineer in chief', although his salary at £400 per year was lower than that of Bush, who was paid £700 as a reflection of his responsibility for delivering Brunel's railway.

Design work was done by Brunel at Duke Street, but he expected Bush to 'carry out in detail the general principles which I may lay down',[22] a situation which made for an uneasy relationship between them, since while making important decisions on the ground Bush was still operating in a subordinate position to Brunel, who only made visits to South Wales when other commitments permitted.

The most notable feature of the railway was its gauge; Brunel decided to treat the Taff Vale differently and did not consider it as a feeder to any planned broad gauge line that might eventually be built in South Wales but instead as a standalone mineral railway that did not require fast passenger trains. This, along with the sharp curves and gradients encountered on the route meant that in 1836 he recommended to Directors that the gauge of the railway should be 4 feet 8 ½ in. This may have been a pragmatic decision at the time, but later became something opponents of the broad gauge would use against him. Following the valley of the River Taff for most of its twenty-four mile route, Brunel's line provided few engineering challenges and it was opened first from Cardiff to Abercynon in October 1840, and on to Merthyr by April the following year.[23]

Two further developments in Brunel's already hectic life took place in 1836. The engineer had noted in his Boxing Day journal entry the previous year 'Mrs B. – I foresee one thing – this time 12 months I shall be a married man. How will that be? Will it make me happier?' With better prospects Brunel's thoughts had turned to marriage; there is little doubt that he enjoyed the company of women and earlier diary entries had recorded a number of attachments, particularly a long relationship with the mysterious figure of Ellen Hulme, of whom only a little is known, since a number of entries in his journals were removed by his family posthumously. His relationship with Ellen had ended in 1829 and three years later he became a regular visitor to the house of John and Elizabeth Horsley following an introduction by his friend Benjamin Hawes. There he met Mary, one of their three daughters; later described by family biographer Lady Celia Noble as 'the family beauty'.[24] While not sharing the artistic talents of her siblings, Mary caught Brunel's eye from the beginning, and after visiting the family on a number of occasions he finally proposed to her in May 1836. The couple were married at Kensington church on 5 July 1836, the wedding marked in Brunel's appointment diary with a large cross! The diary was blank from that day until 23 July as the couple spent a two-week honeymoon touring the mountains of North Wales, returning through the Welsh borders to the West Country. As might be expected, work was never far away from Brunel's thoughts, even during his honeymoon, and he met Charles Saunders at Cheltenham during his absence to discuss progress and review urgent correspondence.

A second development was the result of a meeting of GWR Directors held in October 1835 at the Radley Hotel in Blackfriars at which Brunel was a guest. Thomas Guppy later recalled that a 'festive entertainment' took place after the meeting and that the after-dinner discussion turned to the enormous length of the railway to be constructed, which was at that point the longest yet planned in England. Brunel is said to have told the guests 'Why not make it longer, and have a steamboat to go from Bristol to New York, and call it the Great Western?'[25] In his 1870 biography of his father, Isambard Brunel wrote that the suggestion was considered a joke by many of the diners, but Guppy and Brunel continued their discussion late into the night, with the result that following further discussions with three other GWR Directors, a small committee was set up to transform Brunel's idea into reality.[26] The group, which would form the nucleus of a steamship company, was completed with the recruitment of Captain Christopher Claxton, an experienced naval man, already known to Brunel and Guppy.

Given the immensity of the task facing Brunel in terms of the building of his new Great Western Railway and the other projects he was engaged in, it is difficult to imagine how he could have even contemplated adding such a major task as masterminding the construction of a new transatlantic steamship, as at the time he and Guppy had been developing their ideas for the SS *Great Western* hardly a spade had been turned on his new line. Having travelled the length and breadth of the route during the surveys, he had, as Rolt notes, seen his iron road in his 'mind's eye', but as he wrote when work was about to begin on the GWR, 'I want tools.'[27] These 'tools' required an organisation that would enable him and his band of long-suffering assistants to acquire land, properly lay out the route, and finally construct the line, negotiating with landowners, contractors and the many other individuals and organisations to be encountered as the railway cut a swath across the landscape.

The construction of Brunel's new railway was organised into two separate sections of line, the London Division, which extended from Paddington to Reading, Swindon and Chippenham, and a Bristol Division covering the route from Bristol to Bath continuing through to Box. Each division was managed by a number of 'Resident Engineers', Brunel's 'eyes and ears' when he was engaged elsewhere. In the case of the Bristol or Western Division, G. E. Frere, G. T. Clark and T. E. Marsh were Resident Engineers, while at the London or Eastern end of the line T.A. Bertram, John Hammond and Robert Brereton undertook the work. A separate Assistant Engineer, W. Glennie, was also subsequently provided for the biggest engineering challenge on the railway, the Box Tunnel. Under all these engineers lay a further layer of assistants whose names are scattered throughout the enormous amount of correspondence generated by Brunel and his Duke Street staff during the five years it took to build the railway.

At the centre of it all was the 'Chief'. Brunel, when not on the road checking on the progress of work, worked from his Duke Street office masterminding the design of the railway and ensuring that his vision for the line was delivered by the team employed under him. It seems likely that he did this by first sketching out his ideas for bridges, stations, tunnels, track layouts and all manner of other features of the railway and then sending these to his draughtsmen who would transform these into worked-up full engineering drawings, which he would review and finally sign off. Brunel's sketchbooks, now preserved at the Brunel Institute of the SS Great Britain Trust, reveal the incredible range of information required to build the line, from larger, beautifully drawn sketch elevations of stations like Bristol Temple Meads and Paddington to more technical details of roofs and foundations. The sketches reveal both Brunel's artistic skills and his eye for detail. No aspect of the railway was missed, even down to lamp posts or fencing.[28]

Many finished engineering drawings are now in the care of the Network Rail archive at York and remain a valuable resource for engineers maintaining the railway today.[29] They could, as Angus Buchanan noted, be considered as works of art in their own right and although they bear Brunel's signature, it is probable that a good number were the work of one of the Duke Street draughtsmen, since it is unlikely that he would have had time to do more than check them before signing them off. Brunel's sketchbooks are full of mathematical calculations and other technical information relating to the bridges, tunnels and other structures he designed; in addition, he maintained 'General Calculation Books' that recorded highly technical information such as loading, stresses and other important information and he kept books of 'Facts' that detail experiments on cast iron, timber and a variety of notes about other railway projects. Research undertaken at the Brunel Institute has revealed that the books do contain some mathematical errors, no doubt some may have been a result of the long and unsocial hours Brunel spent at his desk but they may also show that Brunel did have some gaps in his knowledge; Buchanan notes that a number of his assistants did help him in checking calculations on occasions.

Brunel's completed drawings and calculations were then provided to the large number of contractors whose task it was to actually build the railway. The construction of the line was divided into a series of individual contracts, each numbered according to a section of line. Contracts could be for larger stretches of line that might include cuttings, embankments, smaller under or overbridges and buildings, or could be separate, more substantial arrangements for larger structures such as tunnels or bridges. The two Eastern and Western Divisions were further divided into smaller subdivisions for the purposes of contracts, each being identified by the first letter of its location, for example L 'London', R 'Reading' C 'Chippenham and so on. Individual contracts were then numbered to enable the performance of contractors and payments to be properly administered; as an example, the contract for the excavation of the cutting at Sonning was 8L, although as more difficulties were encountered as work on this progressed, the contract was subdivided into three further separate agreements.[30]

Not content with managing all aspects of the design process, unlike contemporaries like Joseph Locke and Robert Stephenson, Brunel also involved himself with the minutiae of the contract process rather than leave it to subordinate staff to administer. This approach, 'resembling that of the Spanish monarchy rather than that of a constitutional government', would ultimately affect both the speed of the works and his long-term health and wellbeing.

F.R. Conder recorded that the ideal method for letting contracts was first to accurately survey, map, level and section the line, the chosen route having been first 'set out' with pegs or flags.[31] Richardson's journal for 1836 has a number of entries describing the work needed to follow up Brunel's preliminary survey by proving the ground and 'pegging out' the route. Based at the White Hart Inn at Brislington near Bristol, the work was hard with long days; while 'Mr B' spent much time at Duke Street, he also regularly appeared in his chaise to check on progress. On 15 March Richardson was at work plotting levels at Keynsham when Brunel arrived, but the following day he had travelled back overnight to London from where Richardson recorded he had then sent copies of plans to him.[32]

Plans and drawings were then produced not only of the line, but individual features such as bridges, culverts, gates, crossings, buildings and permanent way from detailed work done on the ground. A complete set of these drawings, along with a specification prepared in tandem, was then made available for inspection by contractors to enable them to submit a tender for the work based on a lump sum for the whole contract. An example of this process was reproduced in an advertisement printed in the *Windsor & Eton Express* on 22 June 1839 which announced that the Directors of the GWR 'would meet at their office on 9 July…to receive tenders from such persons as may be willing to offer for erecting the station at Reading'. It was also noted that the Directors 'would not bind themselves to accept the lowest tender' although given the ever growing cost of Brunel's new railway, the latter was unlikely to be a regular occurrence.[33] The contractor was responsible for calculating the costs of earthworks and the amount of soil or rock to be moved based on figures supplied by Brunel, with the proviso that the company was not held responsible for the accuracy of this estimate, which meant that contractors bore all the risk. A surviving tender submitted by the contractor Bedborough in March 1838 for contract 4R covering 'Earthworks, bridges and masonry from the bridge over the Thames at Moulsford to the road, No. 16, Parish of Dudcot' (Didcot) illustrates the type of information required by Brunel and the GWR. The total value of the tender was £59,570 and Bedborough included a schedule of prices which included brickwork, all excavations and scaffolding of £1 10 shillings and sixpence per cubic yard, and estimated costs of earthworks that meant the digging and movement of chalk and sand would cost one shilling and fourpence per cubic yard.[34]

When estimates proved inaccurate, even if the fault was that of Brunel's staff who had originally miscalculated quantities or dimensions, his immoveable position was that he and the company were always right and he was the final arbiter of any dispute. Few contractors who incurred additional costs due to errors or unforeseen circumstances ever managed to obtain more than the sum they had originally tendered for, even if the dispute went to arbitration. Brunel soon earned a reputation for treating his contractors badly, making unreasonable demands on them, and varying specifications where details had been left to his discretion to gain an advantage and keep costs down. Brunel's letterbooks are full of snappish correspondence with contractors, making additional demands of them and listing complaints about both the quality of their work and the speed at which it was proceeding, the latter often being the fault of Brunel's office, who struggled to keep up with the huge volume of work involved and were sometimes slow at providing extra information for contractors. Brunel was also notorious for the slow speed at which he settled bills, a trait which may have helped the cash-strapped railway, but one which pushed many contractors to the point of bankruptcy. Even the relatively well run firm of Grisell & Peto was forced to borrow £100,000 in 1838 to pay their men when Brunel failed to promptly settle a large outstanding bill for work done on the Hanwell viaduct which, when interest was added, meant they made ultimately little or no profit on their contract. The result of such behaviour was that larger firms became reluctant to bid for work on the GWR and smaller and less able contractors were employed instead. Some of these failed to meet Brunel's notoriously high standards, which then required more attention and time from him and his assistants to ensure they completed the contracted work, proving something of a false economy in time and effort.

Brunel's insistence on maintaining ultimate responsibility for everything inevitably took its toll; those working under him were reluctant to act without his authorisation fearing a tongue lashing if they exceeded their remit, resulting in more correspondence from them asking for permission or clarification from the boss. They regarded themselves as 'less the officers of the company as the channel of the will of Mr. Brunel' as Conder recalled.[35] Responsibility for design matters and contracts was by no means the end of his work; he also answered other correspondence, met with landowners and investors, and appeared in Parliament when required. Demonstrating seemingly little self-awareness of a situation that was clearly of his own making, writing to Saunders in 1837 Brunel bemoaned his situation with more than a little self-pity:

> I have cut myself off from the help usually received from assistants...No-one can fill up the details – I am obliged to do all myself and the quantity of writing in instructions alone takes 4 or 5 hours a day and as invention is something like a spring of water – limited – I fear I sometimes pump myself dry and remain for an hour or so utterly stupid.[36]

As the engineer in charge, Brunel was required to give a report at the regular half-yearly shareholders meetings for the Great Western Railway and also attended meetings of the two separate committees of Directors in Bristol and London who oversaw the project, a somewhat clumsy arrangement that continued until 1843, when a single board was finally created. These committees regularly reviewed progress on the construction of the line, approved designs and most importantly monitored costs and contracts. A further level of supervision came in the form of a 'Progress of Works' subcommittee and the minutes of these reveal the level of detail involved in the construction process and Brunel's close supervision of all aspects of the work, from design decisions and contracts to more mundane concerns. At a meeting held on 30 October 1838 a letter was read to the Bristol Progress committee from the 'Vicar, Curate, Lecturer and Church Warden of Brislington respecting the employment of Workmen on the line on Sundays'. Brunel was asked that the Resident Engineer George Frere 'be requested immediately to report to this committee as to the facts alleged'. Given the strong Sabbatarian feeling still prevalent in society at the time, the company was clearly anxious not to provoke further opposition to the railway at a time when its presence in the area was not welcomed by all.[37]

In his letter to Frere written early in September 1835 asking him to get on with work at the Bristol end of the line, Brunel had concluded by saying that 'we shall have our flags flying over the Brent valley tomorrow'[38] and it was there, close to London that the first contract let by the company and the first construction work took place. Around seven miles from Paddington Brunel's railway was to run through Hanwell across two embankments crossing the valley on a substantial viaduct sixty-five feet above the River Brent. The contract for what became known as the Wharncliffe Viaduct was agreed with contractors Grissell & Peto, following the submission of a tender by them quoting a price of £75,000 for both the foundations and the brickwork. Work began in February 1836 and despite reporting to Directors that the contract was progressing well, Brunel behaved in what would become a familiar fashion, accusing the contractors of poor workmanship and in a June 1836 letter complained about the quality of the London bricks being used. The engineer demanded that Peto's foreman be sacked as he had ignored his 'particular orders' and Brunel concluded that he found 'upon the works such causes for complaint that I feel called upon to take some decisive steps to protect the interests of the Company'. Three days later he sent another peremptory note summoning Grissell & Peto to his office that day as 'work on the Hanwell Viaduct was proceeding as unsatisfactorily as ever'. In the interim, Grissell & Peto had also taken on a contract to build another bridge across the Uxbridge Road along with further earthworks; the 'skew' bridge was constructed of cast iron with a wooden deck and was awkwardly situated over a crossroads.[39] This bridge and the Wharncliffe viaduct were nevertheless completed on time in June 1837, but after being presented with their invoice for £162,000 Brunel petulantly stalled payment, questioning the contractor's figures and not finally agreeing payment until almost two years later. Peto never forgot the way he had been treated, and the company did little other work for the GWR after that.[40]

As the first major structure on the line, Brunel took the opportunity to produce a design for the Wharncliffe viaduct that was bold and graceful. At the time it was built the structure soared across what was then a beauty spot and consisted of eight semi-elliptical arches, each with a span of seventy feet. These arches were supported by seven massive piers, each with two pillars that were Egyptian in character, similar to the proposed style of Brunel's Clifton Bridge, and betraying the influence of his father Marc, who had assisted him in the design of both. The bridge piers were hollow to reduce the weight of the brickwork, while maintaining the strength of the structure.[41]

The massive 900-ft-long structure sat on foundations seventeen feet deep in the treacherous London clay, a formation that would cause Brunel and his assistants a number of problems through the slipping of embankments and bridges in the area; writing a few months after the completion of the viaduct Brunel reported that settlement in the area had been diminished by the deposit of tons of gravel to stabilise the ground. As recognition of the help the company had received from Lord Wharncliffe, the viaduct was given his name. The Baron had been a valuable supporter of the railway scheme during Parliamentary scrutiny of both the 1834 and 1835 bills, and as Chairman of the Lords committee in the 1835 proceedings he had helped guide the bill to a successful conclusion. It seems that the noble lord was somewhat bemused at this gesture and in his letter agreeing to his name being used nevertheless wrote that since he was neither a shareholder nor landowner in the area, the giving of his name 'to any part of the railway will have no apparent reason for it at all intelligible by the public'.

It seemed that after all the difficulties encountered in the Parliamentary stages, it was wiser to get on with the work as fast as possible, even though the cost would ultimately be greater. Pushing out from the capital, hundreds of workmen began to stream across the landscape, changing what had for the most part been countryside and farmland. The navvies were undoubtedly a tough bunch. With little or no mechanical help available to excavate cuttings and tunnels or build bridges, embankments and stations, many employed by contractors building the Great Western had helped build canals in earlier years. The writer Samuel Smiles gives a fascinating pen portrait of a navvy, noting that they usually wore a white felt hat with its brim turned up, a velveteen or jean square-tailed coat, a scarlet waistcoat with 'little black spots' and a bright coloured kerchief round a 'Herculean' neck. Corduroy breeches retained by leather belt, tied and buttoned at the knee, completed the outfit, displaying a 'solid calf and foot' encased in strong high-laced boots.[42]

As the pace of work on the new line began to accelerate, the navvies' powers of endurance would be sorely tested, with men often working twelve to sixteen hours a day. Contractors were being harried not only by the ever-impatient Brunel but also his assistants. Writing in 1837 Hammond urged his colleague Archibald: 'By all that's good you must get on faster. You must crowd the men everywhere and work night and day.'[43] The need to progress rapidly also led to complaints from landowners and tenants as workmen took liberties in their rush to complete work; on a number of occasions this was because the GWR had been slow to pay compensation to those affected by the passage of the railway across their property. In January 1837 Henry Collis of Castle Bar Hill complained to the London secretary Charles Saunders that men employed by contractors McIntosh had 'commenced fencing off my land which I feel justified in stopping as I have not been settled with'.[44] A great deal of correspondence from Brunel's office also chronicles a longer dispute over stables belonging to *The Feathers* public house in Ealing. Threatening court action, a solicitor reported that contractors had 'acted very improperly by their abrupt method of attempting to take possession'. Saunders told the contractor that he should not 'let workmen trespass until they hear further from him'. It took three months for Brunel to finally approve payment, during which time little could be done on this section of the work, despite the engineer's habitual complaints to his assistants and contractors about lack of progress.

The correspondence generated by Brunel and his Duke Street office paints a vivid picture of just how much detail was required to actually build his railway. While a huge amount of work had been done when surveys had been undertaken in 1834 and 1835, on the ground there were many problems to be solved. The files are full of letters about snags and hitches encountered on a daily basis; walls that needed to be taken down, permissions not granted and always requests for the additional information required by contractors to clarify the sometimes poor quality of information supplied to them in tender documentation. It seems that Brunel's railway surveyors were struggling to keep up, as many requests were for measurement, levels and quantities. The hard-pressed Hammond, again writing to the possibly indolent but probably overstretched Archibald, told him that he had been forced to answer an outstanding letter regarding property close to the line and 'would have done it long ago if you had listened to my repeated request to have all the land measured'.[45].

Benjamin Shaw the Company Chairman had reported to shareholders in August 1836 that the whole of the line had been 'set out', adding that work had commenced 'at all principal points' and highlighting the progress that had been made between Bristol and Bath, which meant, he noted, that the line between the two cities would be finished by February 1838.[46] Trial shafts had been sunk at Box tunnel, described as the 'principal work' on the line; it was noted that the results of these trial shafts had not revealed any difficulties, meaning that the work could be completed within three years, a rather optimistic estimate as events would subsequently confirm. It was also reported that contracts had been laid for the section between Acton and Reading and that work had begun; closer to London the line to Maidenhead should be complete within a year the Chairman argued, but that to achieve this it would be necessary to complete the additional line from Acton to Paddington 'with the same speed and energy' already demonstrated on other parts of the railway.[47]

This somewhat vague mention of the four-and-a-half-mile link between Acton and Paddington concealed a pressing problem for Brunel caused by the breakdown of negotiations with the

London & Birmingham Railway over a location for the GWR London terminus. While it had been announced that the company intended to go back to Parliament for permission to build the short line, delays in getting agreement from landowners affected meant that the company had left it too late to submit a bill for the next session, making it unlikely they would be able to open the GWR to Maidenhead as planned. Undeterred, the Directors took the unprecedented and risky step of starting to build the Paddington extension without Parliamentary permission. Finally confident that they could count on the support of landowners, tenants and parish authorities, the Board agreed to commence what Charles Gibbs called 'the difficult part of the work between Acton and Paddington'. Gibbs estimated a potential loss of between £25,000 and £50,000 was worth risking if the Act was not ultimately successful but, as events were to prove, the gamble paid off, and the additional Bill was finally passed by Parliament on 3 July 1837. A month later a further report to shareholders noted that although the Bill had been passed only a month previously, the works between Paddington and Maidenhead were 'in such a state of forwardness as to ensure their completion early in October'.

The 'works' actually referred only to the trackbed, earthworks and other civil engineering features and the August 1837 report had also contained a note of warning to shareholders stating that the time necessary to lay 'permanent rails in a manner quite satisfactory to the Engineer' would probably induce the Directors to delay the opening of the railway 'somewhat beyond the period first contemplated', hinting that this might not then take place until November at the earliest. This was the first public indication of yet another innovation Brunel wished to introduce on his new railway; not content with having a wider track gauge, he contemplated a different method of constructing the permanent way. Railways built up to that point had largely utilised wrought iron rails supported in cast iron track chairs that were fixed into stone blocks set in the ballast. Brunel had seen this type of track formation in use on the London & Birmingham and Liverpool & Manchester railways. Having travelled on the latter line and having been singularly unimpressed with the rough ride, Brunel was keen to improve on the design and proposed instead what later became known as the baulk road. Rails were supported on longitudinal timbers, usually

30 feet in length, which were braced with cross-sleepers (known as transoms) at regular intervals. This created a robust and stiff framework which was then fixed into the ballast and held down by beech piles which were driven in and bolted to transoms. When the piles were in place, ballast could then be packed under the timber creating, Brunel argued, a solid foundation for the track. The rail, too, was of a very different profile to that used on other lines being an inverted 'u' shape usually referred to as 'bridge rail'. Brunel himself noted in 1838 that the 'simple application of rails upon longitudinal timbers' was not new and that he believed it was 'the oldest form of railway in England'; this method of track construction had been used in workshops and dockyards for many years and it might well have been the case that his father Marc had suggested it, based on track used in the naval dockyard at Chatham.

An additional complication was that the pine used for the baulk road needed to be treated to protect it from the elements. Until the widespread introduction of creosote in the 1840s it was possible to preserve timber by dipping it into chemical tanks containing mercuric chloride, a patented process called 'Kyanizing'. Brunel was forced to justify the consequences of his novel ideas in a report to Directors in January 1838 arguing that the additional cost of £500 per mile against more conventional track construction would be recouped in a few years through decreased track maintenance costs and claiming that the London & Birmingham were already paying more for their rails and stone blocks.

Whatever the benefits of Brunel's new track system were, more practical considerations meant that in February 1838 the company was forced to announce the 'lamentable' and unavoidable postponement of the opening of the line to Maidenhead. It was noted that the Directors were more than aware of the 'extraordinary and unremitting labors which have been bestowed by the engineer' whose new track system 'required the most vigilant attention and devotion of his time'. A longwinded explanation for the delay followed; much of the timber for Brunel's baulk road ordered in the spring had not arrived until November, and even then, only half the expected quantity was delivered. When the balance did arrive, the company did not have enough canal boats to ship it to their saw pits, close to the line. It also appeared that cutting the timber to the correct size was a more complicated operation than first envisaged and 'practical experiments' had been necessary to perfect the process. Shareholders were also told that tracklaying had been further delayed by defective and leaky tanks for the Kyanising of timbers, which had necessitated their repair and replacement.

Detailed specifications for the manufacture of 3,000 tons of rail then needed for the first stretch of the new railway had been produced by Brunel in the spring of 1837 and sent to various manufacturers in South Wales and the Midlands. The rail should be smooth, sound and 'entirely free of hollows, scales, fins or other defects' and would be manufactured 'to the satisfaction of the Engineer of the Company, who shall have power to reject all such as my be defective in any respect'.[48] The railmaking industry was still very much in its infancy, and as a result manufacturers struggled to roll out bridge rail of a high enough quality to satisfy Brunel; the engineer spent much time in correspondence with companies such as Guest Lewis & Co. of Dowlais in his quest to perfect the manufacturing process. He also sent staff to South Wales to inspect and sign off rail as it was manufactured; after consulting with Brunel at Duke Street on 29 November 1837 Charles Richardson was sent to Wales and spent a number of weeks there before and after Christmas. On 2 December he recorded in his diary that he had 'walked to Ebbw Vale works, to watch rolling &c'. In numerous entries in January 1838, he noted further experiments and approval of rails ready for transport to London, regularly writing to 'Mr B' on the progress or otherwise of both experimental work and manufacturing, before returning to Duke Street on 13 January.[49]

While in the long term Brunel's work with manufacturers would ultimately advance progress in the production of high quality rail for the GWR and the railway industry as a whole, the timing could not have been worse. Spending months perfecting the manufacture of a new type of rail when Directors and Shareholders were very anxious to open even a short section of his new railway and begin to earn much-needed income was not ideal to say the least. It was soon realised that installing the baulk road was more time-consuming than had been originally thought, and that further 'practical experiments' were needed to produce as much 'perfection as possible' in constructing the new permanent way. Attempting to reassure shareholders, the Directors finally argued that the winter was not the most ideal time to open the railway anyway and as a result the delays listed in the report would cause 'little injury' to the long-term prospects of the railway. Brunel had, they maintained, through his 'close and vigilant observations of continued experiments' been able to perfect the 'general result of his plan'; with a new deadline of May 1838 for the completion of the London to Maidenhead line Directors and Shareholders anxiously awaited the results of the engineer's new and pioneering ideas.[50]

4

Storm Clouds

Two years after the passing of the Great Western Railway Bill by Parliament, progress on Brunel's new railway was tortuously slow. The combination of his continued insistence on the close supervision of all aspects of railway activity – including not only the survey and construction of the line but also its promotion and management – and the sheer scale of innovations being incorporated into the new railway took their toll and there was concern from both shareholders and Directors at the lack of progress. The diary of George Henry Gibbs, London businessman and GWR Director, records the anxiety felt by the board, but also their undoubted loyalty to and confidence in Brunel, despite the difficulties faced by the railway. On 4 November 1837, having visited Paddington he reported that he was 'not particularly satisfied with progress made'[1] and a few weeks later gloomily added that that he saw little chance of the railway being opened until February. On 6 January 1838 he reported on a discussion with Saunders who had told him that he 'thought Brunel's calculations as to time very deficient' and that he had 'spoken very seriously to him on the subject'.[2] This discussion may have been following a letter to Saunders written by Brunel in December 1837 after working all night on railway business, where the engineer admitted that in the course of introducing 'a few, really but a few, improvements in the principal parts of the work, I have involved myself in a mass of novelties', conceding that he was doing his utmost 'to lead the Company safely through the temporary difficulties I have got them into'.[3]

Nowhere was this reluctant admission of fault by Brunel better illustrated than in the provision of steam locomotives for the railway. Three days after his conversation with Saunders, Gibbs was at West Drayton, with Brunel once again complaining at the unfinished state of the line but pleased to see that two engines were about to be tested. The demonstration did not get off to the best start when the pointwork from the engine shed proved to be too sharp, derailing the engines two or three times before they were finally able to be moved onto the main line. Once ther,e Gibbs noted they performed beautifully and he and Brunel had a 'very interesting drive'. The locomotives were the *Premier* and *Vulcan* and had been supplied by engine builders Mather Dixon and Tayleur. Both had been delivered to West Drayton by barge after a sea voyage from Liverpool and were the first of a number of locomotives supplied to the railway following approaches to various builders by Brunel in June 1836.

At the second half-yearly shareholders meeting of 1836 the Chairman had reported that the Directors 'had pleasure in stating that several of the most experienced and eminent manufacturers of Locomotive Engines in the North have undertaken to construct new engines' for the railway.[4] Brunel had written a generalised letter of specification asking them to provide him with proposals for engines to work on the GWR. Rather than stick with tried and tested ideas, he a took a characteristically unconventional approach allowing detailed designs of locomotives to be left to the individual companies. 'The particular form and construction of your engines will be left to your judgement,' though materials and workmanship should be 'of the best description'. As always, Brunel maintained the final say, requiring builders to supply him with drawings and specifications before construction began, telling them that he would retain the right to object if the design was 'an experiment not worth the making'.[5] The proposition was not completely open-ended, and Brunel listed a number of 'conditions' that were to be complied with. Engines were required to be able to run at 30mph at a relatively modest piston speed of 280 feet per minute, with boiler pressure set at only 50 pounds per square inch; in addition, six-wheeled locomotives were required to weigh a maximum of 10 ½ tons excluding their tenders.

These requirements were in retrospect conservative, given the grand vision Brunel had envisaged for his new railway, in particular the favourable gradients and 'baulk road' broad gauge track, provided to advance high speed travel. To satisfy Brunel's conditions however, in most cases the locomotive manufacturers were compelled to produce small and underpowered engines that had large driving wheels and small cylinders, producing a number of very strange looking machines that were labelled in later years by railway historians as 'freaks'.

Six companies delivered engines to the Great Western in 1837 and 1838, five answering Brunel's initial letter of enquiry producing engines based on his specifications: the sixth, Robert Stephenson & Co. supplied two locomotives that had fortuitously become available after the (temporary) failure of the New Orleans & Carollton Rail Road in America. These two engines, as yet unnamed, would eventually be called the *North Star* and *Morning Star* when operated by the GWR. The first nineteen locomotives built to Brunel's specification by the five original manufacturers were eventually delivered to the Great Western Railway over the next few years, but most were underpowered and unreliable, steamed poorly and were unsuited to hauling the long-distance trains envisaged by Brunel for his new railway. The best of this motley group eventually required substantial rebuilding at Swindon to enable them to remain in service while some were so unsatisfactory that they were quietly and quickly withdrawn, having run relatively short distances.[6]

There has been much speculation about Brunel's reasoning in the case of these early engines but one cannot help but attribute the problems that his specifications caused to a combination of inexperience in locomotive design and operation and his overwhelming desire to control and manage all aspects of his new railway. Rather than relying on the knowledge of others with more practical experience in locomotive engineering, Brunel yet again adopted a more theoretical approach. A safer option would have been to purchase a number of the already tried and tested Stephenson 'Patentee' design locomotives already being successfully operated on other railways in England and Europe, but since the Stephenson company was already very busy with contracts for other railways it would have probably struggled to deliver the engines needed for the GWR in 1837 or 1838 anyway. It is not known whether Brunel discussed locomotive design in detail with Stephenson, but even if this was the case, it seems unlikely that the ever-practical Stephenson would have recommended the unconventional specification suggested by his friend.

In the event, Brunel's conditions meant that it was the locomotive builders who carried the risk and expense of producing new designs, not the Great Western. While the results of these experiments were to cause Brunel and GWR Directors much pain in the short term, it is probably true that lessons were learned from these failures that would significantly improve the quality of GWR locomotive design in later years. The fact that they had essentially become a test bed for Brunel's theories on locomotive technology was not lost on the manufacturers themselves. The Mather Dixon Company of Liverpool was the least competent of all of those supplying engines to the railway and built six engines for Brunel. *Mercury* survived the longest, in use for just over four years, while *Mars* built in 1838 was initially fitted with enormous 10-foot driving wheels that were soon replaced with smaller 8-foot wheels and after various other modifications and repairs was finally approved for service by the company in April 1840. It ran just over 10,000 miles and was withdrawn only eight months later in December. But in a letter to the company written in October 1840, Mather Dixon, while admitting that the quality of their workmanship had been a cause of 'disappointment' to the Directors of the GWR, argued that they felt aggrieved that 'as a matter of justice, individuals should be made to bear the weight of the losses incurred in making experiments for the benefit of a public company, to carry out a principle originating with themselves'.[7] Most of the defects arose 'from our desire, at Mr Brunel's request, to avoid weight', they added.

None of the engines ordered by Brunel had arrived in August 1837 when the Great Western Railway board made a decisive appointment that would go some way to addressing the impending issues resulting from Brunel's inexperience and desire for experimentation. The twenty-one-year-old Daniel Gooch was recruited as Locomotive Superintendent, a vital role given the imminent opening of the first section of line from London to Maidenhead and growing unease at Brunel's shortcomings as a locomotive engineer. Born in

Bedlington in Northumbria, Gooch spent his early life in an area dominated by collieries, engineering works and foundries and the family knew both Stephensons well. Although young, Gooch had already gained much practical experience in heavy engineering and railways, having worked in South Wales, Scotland and the North East; he had also been employed at the Vulcan Foundry in Lancashire and through his family connections had been employed in Robert Stephenson's Forth Street Works as a draughtsman in the early part of 1836.[8] After leaving Newcastle to work elsewhere later that year, Gooch eventually found himself out of work and in 1837 wrote to Brunel offering his services as locomotive engineer to the GWR on 20 July. Brunel was in the process of visiting various locomotive manufacturers in the north west and the two men finally met in Manchester on 9 August 1837. Brunel already knew of Gooch's engineering pedigree and connections and offered him the job in the spot. Some weeks away from his twenty-first birthday, Gooch was delighted and later recorded that he had felt that the opportunity would be 'a permanent thing in which, by attention and perseverance, I might hope to get on'.[9] The two men could not have been more different; Brunel was flamboyant, showy, creative and comfortable in society, Gooch austere, practical and almost puritanical, although no less ambitious or keen to succeed financially. While Brunel thought nothing of parties and entertainments at his Duke Street home or elsewhere, Gooch was much less sociable; after attending his first metropolitan party at the Horsley house in January 1838, he left 'disgusted with London parties' and made a note in his memo book 'never to go to another one'.[10]

Gooch quickly travelled south to join the GWR on 18 August 1837, beginning work by putting together plans for locomotive facilities at Paddington and Maidenhead. He also began the salutary task of visiting locomotive manufacturers to see what progress had been made on the engines ordered to Brunel's specifications. One imagines that Gooch must have wondered what kind of situation he had placed himself in when he began to see just how poor the engines were. 'I felt very uneasy about the working of these machines, feeling sure that they would have enough to do to drive themselves along the road.'[11] Gooch was a strong supporter of Brunel's broad gauge but the design of his engineer's locomotives were 'as bad as they could be' he gloomily later recalled; he had made his mind up to help Brunel deliver his experiments 'and therefore said nothing, but certainly dreaded the result'.

Gooch was provided with a horse and gig, a more modest means of transport than that used by his master, and he moved to West Drayton, the site of an engine shed and the place where a number of the first engines used on the railway were delivered. Despite the important role he would play, Gooch's salary was only £300 per annum, about the same as one of Brunel's Resident Engineers, and when overseeing the unloading of new locomotives he soon found himself on the end of a rebuke the like of which would have been very familiar to his hard-pressed colleagues. Brunel had sent elaborate instructions and sketches of how the engines might be safely unloaded, which, Gooch wrote, 'were of no use in reality, as there was no difficulty in the work'. Without waiting for Brunel or his assistant Hammond, Gooch managed the safe unloading of *Premier* and *Vulcan* but earned a 'scolding' in the process, although unlike many of his peers Gooch responded to Brunel with a strong reply questioning whether Hammond should have any responsibility over locomotive department matters.

Brunel's insistence on being personally involved in every aspect of railway business had a sombre and fatal postscript a few weeks later when he was at West Drayton to supervise the unloading of the first of the two engines purchased from the Stephenson Company, the *North Star*. A set of sheer legs was used to lift the engine from the barge but when a safety rope was let loose too soon the sheer legs collapsed, killing a workman and narrowly missing Brunel. Gooch, away on other business that day, resisted the temptation to tell Brunel 'I told you so,' but wrote later that 'but for the loss of a man's life I rather rejoiced.'[12]

The two Stephenson engines had become available to the Great Western Railway in the summer of 1837; originally intended for use on an extension of the New Orleans & Carrolltown Railway in Louisiana, they were typical 'Patentee' designs with a 2-2-2 wheel arrangement but had been constructed to run on 5 ft 6 in gauge track. With the cancellation of the order, the Stephenson Company had been left with two locomotives that could not easily be sold or adapted for use on any British railway and Brunel's interest in them would have been something of a relief. Brunel, perhaps already nervous at some of the weird and wonderful contraptions being built to his specifications, visited Newcastle and after seeing the engines placed an urgent order for them with the proviso that they be modified to run on his 7 ft gauge rails. Writing to confirm the order in July 1837, Brunel requested that the work be completed in three months and that tenders and an additional pair of driving wheels for each engine be provided at the same time; later records note that each engine cost £2,475 including tender.[13]

There is little doubt that the acquisition of these two engines would have been a real relief to Gooch; they were of a tried and tested design that with his experience of working for Stephenson he would have been very familiar with. Weighing in at over eighteen tons, both were more powerful and heavier than any of the Brunel designs and were certainly a match for anything then hauling trains on railways in England.[14] The *North Star* was first steamed at West Drayton on 15 January 1838, an event celebrated by an alcohol-fuelled dinner the following day, which was spoiled, the ever abstemious Gooch recorded, by some of the guests amusing themselves 'by dancing an Irish War dance on our hats', piled up in the corner of the room.

While the completion of the line from Paddington to Maidenhead was of paramount importance, in February 1838 it was reported that work continued elsewhere along the route, particularly at Sonning where a great cutting was being excavated, and at the west end of the line where work had begun between Chippenham and Box Tunnel.[15] A technical report by Brunel on the method of laying his new baulk road track was included in the report of the February board meeting and he noted that rain and frost had slowed progress but that the experiments carried out had been a success and that the timbers had stood 'most satisfactorily'. The Directors remained loyal to their engineer and by 9 March Gibbs recorded that he was 'well satisfied with the progress of the line' and that he saw no reason to doubt that the line to Maidenhead could be completed in May,[16] although Brunel's assistant Hammond was more cautious saying that it could be opened comfortably in June. By the end of the month Gibbs gloomily reported that what he had seen walking along the line from Acton 'did not give me much confidence as to opening in May'.[17] Even without Brunel's innovations the railway builders faced many problems as they struck out eastwards from the capital. As an example, contract 4L awarded to Grisell & Peto covered a section of line from the Paddington canal to the Iver Road where the navvies encountered many streams and shingle as they created earthworks, requiring the construction of coffer dams at some expense to divert the water.

Brunel's original costing for the railway was proving to be a huge underestimate given the changes and innovations already introduced. Gibbs noted that on 23 February 1838 Brunel and Saunders were 'up all night going over the estimates', a task that clearly provided little comfort to the Board as he suggested that at the forthcoming shareholders meeting the reporting of costs should be as vague as possible so that, without actually practising a deception, the public should not then know the 'real state of the case'. Gibbs wrote that from Saunders' 'manner and expressions' he feared that the railway would require a further £666,666, meaning that the total cost of the line would then be around £4 million, which he hoped would be enough.[18] By the time the railway opened throughout in 1841, costs had risen to over £6,000,000.

On 31 March 1838, Brunel took time away from the anxieties of managing contractors, navvies and assistants to travel on the SS *Great Western* that had been fitted out in London after her launch in Bristol in July 1837. The ship was due to travel from the Thames back to Bristol but off the coast of Essex lagging around the base of the funnel caught fire. As members of the crew including Captain Claxton attempted to fight the fire in the engine room, Brunel tried to descend to investigate. The ladder he stepped on broke, and he plunged twenty feet into the smoky engine room; once again luck was with him and his fall was broken by the unfortunate Claxton, who was standing at the bottom of the ladder. Unconscious and face down in several inches of water, Brunel was picked up by the stunned but unharmed Claxton and hauled out of the chaos by rope, surviving yet another potentially deadly accident. He spent the next three weeks in bed, but it was not long before he was back at work, writing letters and instructions about both his steamship, which set sail for New York on its maiden voyage on 8 April, and also progress on the railway.

Gibbs met Brunel at Paddington on 21 April and found him 'better than I had expected', although in a letter to Thomas Guppy some weeks later the engineer admitted that he was not particularly well 'in body or mind' telling his old friend that he was lame in the left foot and that his back was weak.[19] Huge efforts were now made to complete the opening stretch of the railway to enable an opening by the end of May; at the centre of the tumult was Brunel who had written some months before:

> If I ever go mad I shall have the ghost of the opening of a railway walking before me, or rather standing in front...and when it steps forward a little swarm of devils in the shape of leaking pickle tanks – uncut timber – half-finished station houses – sinking embankments...unfinished drawings and sketches of details will quietly and quite as a matter of course and as if I ought to have expected it, lift up my ghost and put him a little further off than before.[20]

On 28 May, Gibbs was finally able to record in his diary that 200 tickets had been issued on that day for the opening of the railway scheduled on 4 June; in advance of this the company held a celebratory event on May 31, an inaugural run from Paddington to Maidenhead for Directors, their friends, Members of Parliament and 'such other people as, for various reasons were deemed worthy of a place'. The *Birmingham Journal* for 9 June 1838 reported that the invited guests who had begun to arrive at ten o'clock included 'several ladies'. They had a little time to wait at the half-finished Paddington before the six-coach train left at eleven o' clock to the sound of 'cheers and acclamation of a vast concourse of spectators'.[21] The special was hauled by the *North Star* and made its way through Hanwell, Southall and West Drayton, then still a largely rural 'wooded and fertile' landscape, reaching Maidenhead in forty-nine minutes at an average speed of twenty-eight miles per hour. The sedate pace was because the driver had been instructed to take care not to go too fast, as parts of the line were still not completely finished.

Maidenhead station, the first temporary western terminus, was situated on the west side of Dumb Bell bridge where the line crossed the busy Bath Road and was actually closer to Taplow than the town of Maidenhead itself.[22] The location had been deliberately chosen by Brunel and the Company to steal valuable coach traffic travelling along the Bath Road, since Maidenhead was then reputed to have more carriages passing through it daily than anywhere else in England. In 1834 the GWR had employed a man named Dinorben Hughes to make a record of all the traffic passing through the town. In a two-week period that March he noted a huge variety of traffic including 776 carriages drawn by four horses, 2,230 horse-drawn vans and wagons and a large number of specialised carts laden with coal, grain, hay, timber and market produce, not to mention livestock including 2,803 sheep and 102 cattle.[23] Such was the revolution brought by the coming of the railway that in a few short years most of this traffic would be carried on Brunel's GWR. The station buildings were modest timber structures perched on the top of the new railway embankment, accessed by a long flight of stairs; each had small towers, one equipped with a clock, set to Greenwich Mean Time, which Great Western trains ran to as opposed to local time.[24]

After inspecting the station, the invited guests boarded the train once again and returned east to the nearby Salt Hill, outside Slough, where a 'very excellent and extensive repast was provided by the Directors to which all parties appeared to do simple justice', *The Times* reported. The celebration was held in a large marquee pitched in the pleasure grounds the newspaper added, 'very elegantly decorated' with laurels, evergreens and banners. Speeches followed the dinner and after the usual toasts to the Queen and Parliament, Josiah Guest MP for Merthyr Tydfil and first Chairman of the Taff Vale Railway proposed the health of Mr Brunel 'to whose exertions and skill we are so much indebted'. Mr Brunel, he added, had 'stepped out of the ordinary track' and he believed 'no other man would have had the courage to do so.' It was reported later that Brunel returned the toast conveying a feeling of quiet confidence in work, adding that he had 'no personal vanity to gratify in the success of the experiment and 'looked to posterity for his reward, if the attempt merited it'.[25] Perhaps aware of the difficulties his new ideas had already placed the company in, the engineer was careful to thank the Directors 'for the confidence they had placed in him, and he was happy to find that the result had justified that confidence'. Further toasts were drunk to the Chairmen and Directors of the Bristol & Exeter, Cheltenham & Great Western Union, and Taff Vale railways, all eventually to become part of the GWR, and all already having Isambard Kingdom Brunel as their engineer.

There was also much celebration of the success of one of Brunel's other projects, the SS *Great Western*, which was shortly to reach New York after its maiden voyage. To what *The Times* recorded as 'immense cheering', Robert Bright, Chairman of the Bristol committee, said that although progress at the Bristol end of the line to that point had not been great, 'they had not altogether been idle in Bristol for they had laid a railroad of 3,000 miles in continuance of the Great Western and their Railway was laid on the ocean.' A great tide of passengers bound for the United States should, he concluded, be through the Great Western Railway. As the celebration concluded, a further toast was offered to the Directors of the Great Western Steamship Company and was responded to by Thomas Guppy. It may be that Brunel's loyal friend and supporter had himself imbibed a little too much champagne as on the speedier journey back to Paddington it was noted that he had walked over the roofs of the carriages from one end of the train to the other as it barrelled along at full speed, a hair-raising stunt even Brunel would not risk.[26]

The first stretch of railway finally opened to the public on 4 June 1838. Because it was a bank holiday Whit Monday, it was a day *The Times* reported 'not injudiciously chosen for the commencement of operations', and proved to be busy, the company carrying 1,479 passengers and taking its first £226 in income from tickets. Trains left from Paddington and Maidenhead hauled by the locomotives *Aeolus* and *Apollo* at 8am and the company ran a simple timetable of 4 trains each way in the morning, and the same from 4pm onwards later in the day. It was reported that many people travelled merely to experience this new phenomenon, while others went to be ready for the Eton 'Montem', an annual event held by the college, ironic given the college authorities opposition to the railway. *The Times* had noted that 'the pace yesterday was good' and that the journey had been achieved in one hour five minutes, but Gibbs and Gooch were less happy; Gibbs had travelled on the first train out of Paddington and wrote that he was disappointed with a journey time of one hour twenty minutes. This equated to an average speed of only 25 ½ miles per hour, not what had been promised. There was much public enthusiasm at the start of operation, with more than 100,000 passengers travelling on the railway between June and August. On 14 June the railway carried a large number of passengers from the capital to attend the Gold Cup race at Ascot. Horse racing was a national obsession in the nineteenth century and the GWR and other railways would generate much income from race traffic, not only to Ascot but at other tracks such as Newbury, Cheltenham and Bath in future years.

Within weeks of opening, though, it became apparent that the smooth ride for passengers promised by Brunel was not apparent as the track settled and the pinning down of the broad gauge formation with timber piles was not working, in fact having the opposite effect, instead transforming the line into a switchback producing what one observer called a 'see-saw' motion. Gibbs noted on 21 June that 'the road is evidently deteriorating under the pressure of trains.'[27] The rough ride was intensified by the design and poor build quality of the carriages, some of which had inadequate springing and more seriously, wheel tyres of differing thicknesses. It will also not come as a surprise to learn that many of the locomotives built to Brunel's specifications broke down or simply did not have enough steam to pull their trains, with Gooch spending many nights at the Paddington depot working tirelessly to provide enough motive power for the following day's service.

Considerable effort was made to alleviate what was becoming a difficult situation, not only for the railway both through loss of reputation and a fall in share prices but also for the future prospects of its engineer. The teething problems being encountered had not gone unnoticed by Brunel's opponents, who were quick to argue that his love of novelty and extravagant claims about his new railway had not been justified. Brunel had faced particular criticism for some considerable period from a group of shareholders based largely in the North of England nicknamed the 'Liverpool Party'; Gibbs had recorded in his diary for New Year's Day 1838 that a letter from shareholders in Liverpool had been received stating that they had lost their confidence in the company, and there are further entries which seem to indicate that northern shareholders were intent on not only destroying Brunel's reputation but gaining more control of the management of the Great Western Railway itself.

While the poor performance of the railway was the initial focus of their discontent, the whole question of the broad gauge and Brunel's overall competence soon became apparent. In July, Gibbs noted that the Liverpool faction were 'bent on crushing

Brunel' and were using the fall in share prices to help achieve this; one of his strongest supporters, Gibbs nevertheless recognised his engineer's weaknesses, arguing that despite all his talent 'he has shown himself deficient...in general arrangement,' referring to Brunel's unerring obsession with overall control and lack of delegation. 'There have been too many mistakes; too much doing and undoing,' he added.[28] The loyal Gibbs mused that it seemed too early to abandon the work after only a few weeks and that much of the outcry was aimed at Brunel personally rather than the railway.

The first proper skirmish in what became a career-defining period for Brunel came on 15 August 1838 at the sixth half-yearly meeting of the proprietors. A large room in Merchants Hall in Bristol was packed with shareholders from Bristol, London, Liverpool and the North of England. Such was the importance of the event that the *Bristol Mercury* newspaper printed a special supplement to its normal weekly edition that gave full details of the reports delivered by the Company Chairman and Brunel, as well as an account of the tetchy discussion that followed. Between 400 and 500 people were thought to have attended, the correspondent noting that many people were forced to listen to proceedings from various ante rooms as there were no seats to be had in the main hall.

The Company Chairman William Sims told the meeting that in the first ten weeks of operation the railway had earned more than £15,000 while carrying 100,222 passengers, but not much else he said brought comfort to anxious investors in the line. He stated that 'the increased gauge of the rails, and the condition of the permanent way, have been sources of mixed anxiety and satisfaction in the minds of the Directors,'[29] and announced that three of the most eminent professional men in the country had been approached to inspect the line, investigate its working and produce reports, a suggestion made to the board by Brunel a month earlier as criticism of his methods began to increase. Two of the three suggested candidates, George Stephenson and Sir James Walker, President of the Institution of Civil Engineers, had perhaps already wisely declined, the latter giving his reason that it was his impression the matter would be 'controversially conducted'. Only Nicholas Wood, 'a man of high character and extensive experience of the working of railways', had agreed the Chairman told the assembly, but he had yet to begin work. If all this was not enough, Sims left the worst news to the end of his speech, announcing that the Directors now thought the cost of construction would be more than four million pounds.

After this bombshell, Brunel rose to deliver a characteristically bullish defence of his plans; while there had been 'numerous difficulties' they had either been overcome or would gradually and successively diminish, he argued. What followed was a stout defence of all aspects of his new line. He referred back to his earlier report on his reasons for adopting the wider gauge and repeated his point of view concluding that 'every argument here adduced, and every calculation made would tend to the adoption of the 7 ft gauge.' Any increased construction or land costs as a result of the broad gauge were exaggerated or minimal he noted; downplaying the impact of his new scheme he implied that the dimensions for bridges and other structures then being built were the ones he had originally specified in 1834.

Moving on to the difficulties faced by Gooch and the locomotive department, Brunel was unsurprisingly keen to distance himself from the performance of the engines he had commissioned from various manufacturers. He had left them 'unfettered' to produce the best engines they could to his specifications but conceded that even if these engines 'had not been found effective, at least it must be admitted that the best and most liberal means have been adopted to procure them, but I am far from making such an admission', adding defiantly that recent trials had destroyed the arguments of 'alarmists' who had been so critical.

Finishing with some observations about the poor performance of his permanent way, Isambard maintained that his system was not as new as critics had implied, describing it as 'an old system recently revived' and reminding the meeting that hundreds of miles of track had successfully been laid in this way across America and parts of England. It was not the track formation but the way in which it had been laid, and the defective use of piles that had led to the rough ride experienced by passengers. The problems were not his fault but those of the contractors who installed the track without 'due care in the execution'. He had been 'most unfortunately prevented by a serious accident from even seeing the work until almost the opening day', he continued, when he ought to have personally supervised the operation. Brunel offered a lukewarm defence of his assistants insisting that the work had not been neglected, but 'in such a case, new work cannot be properly directed except under the eye of the master,' he pompously concluded.

Once these two lengthy reports had been delivered there was a protracted and often vociferous debate.[30] There were many in the room who, although worried about rising construction costs, were supportive of the railway; Mr Hinton Castle was typical, telling the meeting that he had not been prepared for so large an outlay but having read the accounts and heard the report of 'their able and talented engineer' he was content. Others were less forgiving; a Liverpool shareholder said that the reports were clouded with so many doubts and difficulties that it was difficult to come to conclusions. A request was made to appoint another engineer to work with Brunel and for more representation on the board from the northern shareholders, but neither went to a vote. Another northern shareholder, Mr Hoyes, sarcastically pointed out the 'magnificent' income of £200 per day and the expenditure of £300 per day should not be forgotten. Hoyes also told the shareholders that he felt that more stress had been laid on the width of the track gauge than the way the track had been laid, and that he was equally opposed to both Brunel's novelties; he asked whether the company intended to abandon Brunel's 'mode of laying rails'. The Chairman gave a rather equivocal answer and the engineer continued to defend his arrangements arguing that the railway 'should not be obstinate in adhering to the method of effecting it'. Pressing on, having not really had a satisfactory answer, Hoyes claimed he was not personally hostile towards Brunel, but shareholders had 'been deceived over the time, in the cost, and in the execution of the works'. Many critics would have been more patient over the delays in opening the line he concluded, if the track had been a 'credible one'. He reminded the meeting about the extraordinary alterations that had from time to time been made by Mr Brunel in his 'mode of proceeding'. The shareholders were not 'alarmist' he argued, adding darkly that if the GWR was to continue to be assailed by disaster after disaster, objections would grow.

Hoyes' interventions had been one of many in a long debate and the meeting dragged on for more than six hours, by which time many had drifted away. Speaking late in the proceedings, another shareholder, Dr Carpenter, who, the *Bristol Mercury* reported, had 'made a speech of some length' then objected to business being conducted at such a late hour. The meeting concluded at 7 pm without a formal resolution to remove Brunel and while the Liverpool Party had made many interventions they had seemingly failed to seize the advantage on that occasion. This was by no means the end of the matter and soon after, John Hawkshaw, the engineer of the Manchester & Leeds Railway, was also appointed to produce a report on the Great Western Railway. Throughout the autumn of 1838 both he and Nicholas Wood carried out various experiments on the line. By early October Hawkshaw had completed his work, but Wood's report was not yet complete, although Gibbs described his initial findings as a 'despicable, useless document'.[31] Hawkshaw had little good to say about the broad gauge and noted that any company not adopting the standard gauge was in danger of isolating itself, a claim that would be repeated years later.

Replying to this criticism Brunel admitted that the gauge question was 'undoubtedly an inconvenience', but added that because the GWR was being built in places where 'railways were unknown' and the companies associated with it were a complete system within the areas they served, once his broad gauge network was completed no other lines would be needed, avoiding the problem of areas where two competing gauges might meet. As later commentators noted, for someone so obviously gifted, this argument was strangely short-sighted and difficult to comprehend. It would return to trouble him again as his broad gauge network began to probe north and south into the territory of other rivals in the mid-1840s. While openly critical of the broad gauge itself, Hawkshaw's report said very little about Brunel's track, only stating that the engineer had attempted to do something 'in a more difficult manner, which may be done at least as well in a simple and more economic manner', perhaps a recurring theme in Brunel's engineering career.

Wood's research dragged on while he carried out more research on the merits of both the track and the locomotive fleet. A comparative study of both broad and standard gauge locomotives was undertaken, with tests taking place on the GWR and other railways, with Wood assisted by Brunel's old nemesis Dionysus Lardner. The results highlighted the disappointing performance and coke consumption of Great Western engines, even the seemingly reliable and powerful Stephenson locomotive *North Star*, which Lardner found could haul a load of 82 tons at 33mph, 33 tons at 37 mph, and only sixteen tons at 41 mph. Wood attributed these failings to the greater air resistance provided by larger broad gauge engines, returning to a theme expounded by Lardner during the committee stages of the GWR bill. From these results Wood concluded 'It would not be advisable to attempt an extreme rate of speed and that 35 miles an hour...may be considered the limit of practical speed for passenger trains.'

Nicholas Wood was also highly critical of Brunel's track construction, particularly the use of timber piles, although he did concede that if larger timbers were used then the end results were still better than stone blocks or cross sleepers. He argued that the broad gauge was a more expensive proposition and that converting what had been built would cost £123,976 and the Company could save a further £156,000 by completing the rest of the railway as a standard gauge line. For all but the most loyal Brunel supporters the evidence submitted by Hawkshaw and Wood was damning. On 14 December members of the board suggested that another engineer should be recruited to work with Brunel, with the name of Joseph Locke, Engineer of the Grand Junction Railway, being suggested. Gibbs, Saunders and Charles Russell visited Brunel in his office the same day, and they found him in defiant mood. He knew that evidence was accumulating against him he said, but still 'felt confident in the correctness of his views and was sure that he should have opportunities of proving it'. He would not agree to the idea of working with Locke and would rather resign if that was what the board wanted. Brunel told them that he was convinced that many of Wood's arguments could be refuted and, given a little time, he could do so.[32]

Brunel got the time he needed and discussion of both reports at a shareholders meeting was adjourned until the New Year, despite the misgivings of some board members and criticism from the Liverpool faction, who had spent the last few months agitating and scheming to get more representation on the board. Brunel and his locomotive superintendent Gooch spent much of this time working on improvements to engines, in particular the *North Star*; Gooch recorded in his diary that they had made significant improvements in the steaming of the engines by changing the shape of the blast pipe, the pair spending most of Christmas Day 1838 on the task.[33] Once they had confirmed their findings, he noted that they kept their trials on these matters 'very quiet, intending to spring it as a mine against our opponents once they had committed themselves'.

Examination of both reports could not be delayed further and a special shareholders meeting was convened for 9 January 1839 when the question of the broad gauge should be decided once and for all. The Directors report highlighted that the proprietors should be 'deeply sensible of the disastrous consequences inevitably arising from the continual discussion of the principles acted upon in carrying out the works'. Such was the importance of the debate that the *Railway Times* had reprinted both reports in a special edition in advance of the meeting. Its editorial thundered:

> Mr Hawkshaw's report, as the public have been for some time aware, is entirely condemnatory of Mr BRUNEL'S plans. Mr WOOD'S is, in substance, equally so (somewhat contrary to general expectation), though in a manner less firm and decided. DR LARDNER is but a clever annotator on WOOD. [All three] arrive, at last, at the same memorable conclusion – that nothing worthwhile has been gained, or is ever to be gained, by Mr. Brunel's new-fangled and costly schemes.[34]

The editor continued that he had heard that the 'young man' Brunel was supported by a majority of his Board of Directors. If these things are so, he concluded, then there could only be one way for shareholders to save their Company from ruin, and that was 'to get rid, as speedily as possible, of both Engineer and Directors'.

Given this kind of emotive language it was not a surprise that the proceedings on 9 January were described as 'very stormy' with opinions as 'wide as the poles asunder'. The Directors report made a number of concessions while refuting the conclusions of both Hawkshaw's and Wood's reports. They recommended 'retaining the gauge with the continuous bearings, as most conducive to the interests of the company', but accepted Wood's findings that the use of piles had been responsible for the poor condition of the track, making it too stiff and rigid. The use of piling would be abandoned and a heavier type of rail should be used. Brunel was able to reveal the results of the work Gooch and he had been doing in the previous weeks and disprove much of Lardner's evidence at the same time.[35] *The Railway Times* was less convinced, wondering how Brunel's experiments made with the same engine (the *North Star*) in less than three months could show an improvement of nearly two-thirds in comparison with previous experiments. 'What evidence is there of the truth of this extraordinary – this almost incredible "change of capability"? Mr. BRUNEL says – "I can prove it." Can anybody else? Is the fact taken for granted on Mr. BRUNEL'S authority alone?'[36]

It was inevitable that despite the robust response from the company and its engineer and numerous speeches in support from luminaries including Brunel's friend Charles Babbage, and a 'wretched' display by their opponents, shareholders were not easily convinced. An amendment was tabled noting that 'the reports of Messrs Wood and Hawkshaw contain sufficient evidence that the plans and construction pursued by Mr Brunel are injudicious, expensive and ineffectual ... and therefore should not be proceeded with.' This amendment was defeated by a close margin of 7,790 to 6,145 votes, not a ringing endorsement of the engineer or his broad gauge experiment but for now at least the gauge question had been settled. Gibbs recorded that various members of the Liverpool faction had agreed to end their constant lobbying but that these promises were not to be relied on. The *Railway Times* considered that 'the victory of the Great Western Railway Sesquipedalians [would be] of singularly brief duration.' It also published a letter from Nicholas Wood replying to the report of the GWR Directors, which proved, it argued, that it was 'a triumph based on trick and misrepresentation'. The editor regretted the use such harsh terms but 'no others would sufficiently describe the real character of the proceeding.'[37]

A postscript to this saga was inflicted by the Directors when Daniel Gooch was asked by them to report back on the state of the locomotive fleet as it then stood, without the knowledge of his boss. Following the failure of so many of the engines built and delivered to Brunel's specifications the Directors were, Gooch later recorded, 'very anxious' and he felt placed 'in a great difficulty' as he could only report the facts, which did not reflect well on his master. Gooch nevertheless responded in some detail, concluding that the two Stephenson engines were the only two on which he could rely.

Brunel was predictably upset by this apparent act of disloyalty and sent his locomotive engineer an 'angry letter', although Gooch recorded in his memoir that 'he only shewed it in his letter and was personally most kind and considerate to me.'[38] Regrettably, Brunel's letter does not appear to have survived so we can only guess at its tone and contents, but given all the other worries he had at the time, he may well have been content to leave the locomotive fleet to Gooch, who shortly after began designing 'Firefly' class locomotives based on Stephenson principles that would provide a powerful and reliable fleet of engines that would be the foundation for the GWR motive power department for some years.

Having weathered the storm generated by the Liverpool Party and the broad gauge question, Brunel could now concentrate on the main job in hand, actually completing his railway. Opposition from Eton College remained and it had not gone unnoticed that publicity for the opening of the railway from London to Maidenhead on 4 June 1838 said that trains would run between London, West Drayton, Slough and Maidenhead. Tickets could be obtained at the *Crown Inn*, Slough, the handbill noted, 'subject to there being room in the carriages on the arrival of the train'.[39] The Directors had already hinted that despite the provisos of the GWR Act, they would convey passengers without 'the conveniences of a station' following requests from people in the neighbourhood of Slough.

Local people had also petitioned Eton College to allow the construction of a station, but the Provost remained obdurate and applied for an injunction in the Court of Chancery against what it saw as a challenge to the spirit if not the letter of the law. This was dismissed with costs by the courts and the railway was able to stop trains at Slough, to enable passengers to join trains even if there was not a station at which to do it. Bowing to pressure from all sides, eventually the College accepted the inevitable, and Brunel was finally able to design and commission a station at Slough, which opened in June 1840. The station was one of the engineer's unorthodox 'one-sided' designs consisting of two entirely separate 'arrival' and 'departure' stations, each with their own platforms, offices and refreshment rooms located 50 yards apart and separated by a web of sidings and wagon turntables. It was described as 'a most important station' for the 'many hundreds who resort hither, either going or coming from Windsor'.[40]

As the line pushed along the Thames valley there were two main engineering challenges to be overcome before the railway reached Reading, the largest town in Berkshire. The original Maidenhead station was constructed on the east side of the river and to cross it Brunel created one of his most daring and elegant designs, which was, once again, to bring criticism from some quarters. The Thames was almost 100 yards wide and since it was still in use by barge traffic the river commissioners insisted that neither the main channel nor towpath be obstructed and that sufficient headroom be provided to enable barges and boats to pass safely. Brunel was also reluctant to elevate the level of the track on either side for fear of losing the gentle 1 in 1,320 gradient he had carefully engineered for the line between Paddington and Reading. Aware that he was courting controversy, Brunel wrote in August 1836: 'I have done what I suspect I will be severely criticised for, namely a two-arch bridge designed peculiarly with the regard to the navigation and without consideration of expense on the part of the railway.'[41]

The solution to this puzzle was a bridge consisting of two graceful, semi-elliptical arches of 128 ft with a mid-river pier set on a convenient shoal, with smaller arches on either bank. The bridge was constructed almost entirely in brick and its arches were the flattest and largest then built in this way. Brunel's original foolscap sketch for the bridge survives, the drawing surrounded by a series of arithmetical calculations. Brunel worked closely with the contractor Bedborough who was also responsible for the foundations and embankments on either side, but by March 1837 the contractor, short of money, gave up the contract, and two months later a new contractor, Thomas Chadwick, was recruited to complete the work. By the end of February 1838, the bridge was largely complete; the arches did indeed look flat, and critics were convinced that once the wooden centring supporting and shaping the brickwork was removed the bridge would fall down. In May, Chadwick eased the centring without Brunel's permission and because cement on the eastern arch was not yet properly set, some of the brick courses separated, much to the glee of Brunel's opponents. The engineer explained the circumstances to Directors on 19 June, explaining that he had expected some compression of the arches, but that the present situation was caused entirely by 'the fault of the contractor', and Chadwick had been instructed to make good the work and keep the centring in place for the time being.[42] No doubt there had been some strong words to Chadwick but none seem to have survived in the correspondence; the centring remained in situ over the winter, and was not finally removed until 1840.

The Directors finally had something to celebrate on 1 July 1839 when the next section of line from Maidenhead to Twyford was opened; the Company was keen to push on to Reading as soon as possible, but a different challenge faced the railway east of the town at Sonning. Brunel's original proposal had been to drive a long tunnel through farmland belonging to Robert Palmer MP at Holme Park, even though the railway would have passed almost three-quarters of a mile away from Palmer's mansion – which he did not even live in. All this had been devised to mollify the MP who was a determined opponent of the railway. Eventually the tunnel plan was dropped in favour of a cutting instead.

The contract for this work was let to William Ranger, and it soon became clear that the cutting would be no less challenging than a tunnel. It is highly likely that Brunel was aware that the geology of the area might prove difficult before work started, having had already conversations with his father Marc, no stranger to digging in difficult conditions in the Thames Tunnel. Progress was slow over the winter of 1836 and early 1837 and the works became very waterlogged after constant heavy rain. The gravels and sand encountered in the excavations led to sometimes fatal landslides and for nearly three years hundreds of navvies equipped with wheelbarrows and horse-drawn wagons laboured in terrible conditions in what had become a morass of mud. Accidents were common, and the pages of the *Reading Chronicle* regularly featured reports of fatalities: on 25 April 1839 it was reported that 'a man was killed on the railway at Sonning Hill – the dirt fell on him. A boy was killed in the same place on the 10th of this month. He was run over.' In July five men were badly injured when a barrel of gunpowder intended to shift a vein of hardened clay exploded prematurely. One man 'was raised in the air some distance with the truck he sat on'.[43] Perhaps stung by the stream of casualties, the Directors of the GWR had made a donation of £100 to the Royal Berkshire Hospital, bolstered by an annual subscription of ten guineas, the newspaper added.

Frustrated at the lack of progress, Brunel withheld payment from the contractor, and when Ranger ran short of money he was unable to pay his men, who promptly marched to the nearby town of Reading where eventually only promises from the mayor prevented a major breach of the peace. Brunel removed the unfortunate Ranger from his contracts in the spring of 1838 and the work was divided into smaller contracts. The company would eventually be locked into a Dickensian legal case with Ranger and his estate for years, as efforts were made to recover money they claimed was still owed to them by the GWR. It was hoped that by dividing the contract into smaller parts that the work could be completed in seven months, but this was not to be. Less than two months later Knowles another contractor, working on the west end of the cutting, was also let go having failed to pay his men, resulting in a strike which slowed work again and caused the local population much anxiety as bands of disgruntled navvies roamed around the countryside.

Shareholders were informed that the Company had taken the difficult decision to take over the work themselves, to ensure that the cutting could be finally completed. *The Berkshire Chronicle* reported that 'to make up for the time lost on the G W Railroad near Reading, upwards of 100 extra workmen are now employed, who work at night by fire-light.'[44] Winter storms yet again brought work to a standstill but early in 1839 Brunel's resident engineer reported that 1,220 men and 196 horses were employed at Sonning and that two steam engines had been purchased to assist with the work; there was still 700,000 cubic tons of soil to be excavated it was reported, and this task took much of the year to complete. Travelling by train along the cutting today one is struck as to how deep this massive excavation is, and just how much effort it must have taken to dig it out. On 14 March 1840, Brunel and a group of GWR Directors travelled on a special train from Paddington to Reading, the service hauled by Gooch locomotive the *Evening Star* that completed the journey in just over an hour. They were greeted by cheering spectators on nearby Forbury Hill close to the new station. Several thousand more people gathered on 30 March when the line was formally opened for business. Seventeen trains travelled to and from London that day, signalling a major step forward in the development of the railway, and the day was celebrated with the usual mix of band, crowds and civic receptions.

Despite Brunel's pretensions of building a grand railway on a grand scale, the facilities provided at Reading were far from that. His original proposals for an elegant station were rejected by the ever-careful London committee on 11 April 1839. Gibbs recorded that 'the estimates were far beyond anything we had imagined,' and they were determined not to undertake any unnecessary expense without 'very serious consideration'.[45] Brunel was not pleased and as a result instead produced another 'one-sided' station design in the same vein as Slough. Two separate sets of platforms and associated buildings were situated on the south side of the line linked by a complicated and dangerous track layout that required trains to cross from one line to the other on multiple occasions. The buildings were plain to the point of ugliness and some measure of the disappointment felt by travellers can be gained from the offhand comment in a later railway guide that, while provided with every convenience for the great traffic of so important a place, Reading station was 'by no means as commodious and well adapted to its purpose as those at Didcot and Swindon'.[46]

5

Pushing West

The line to Reading complete, Brunel's contractors pushed further into Berkshire, while at the other end of the railway work continued apace between Bristol and Bath, and also on the most difficult challenge on the whole route, the tunnel at Box. Despite the need to complete the Reading-Bath section as soon as possible, the Directors were acutely aware that the remaining area to be crossed by the railway did not then contain any really large centres of population that might generate significant income for the company, unlike the eastern and western extremities of the line. These arguments had of course already been rehearsed during the first ill-fated 1834 Parliamentary Bill, which had proposed the construction of only those two ends of the route, but the Directors knew that they were now committed to the completion of the whole railway by later legislation. What would help, however, was the success of two associated connected railway schemes that would feed traffic on to the main line, both of which had been noted in the 1835 prospectus. The first was a railway from Cheltenham and Gloucester that would run via Stroud to connect with the GWR main line at Swindon, providing a direct link to the capital. The second was a branch to the university city of Oxford which had been directly promoted and supported by the Great Western in 1836.

As early as 1833, the publication of the prospectus for the GWR and reports of Brunel's plans had prompted discussions within the business community in Cheltenham and following a public meeting in September 1835 it had been agreed to form a company to build a railway from Cheltenham and Swindon via Gloucester, Stroud and the Chalford Valley.[1] Brunel was appointed as its engineer and by the following June the Cheltenham & Great Western Union Railway had been authorised to build the line, along with a branch from Kemble to Cirencester with a capital of £750,000.[2] The initials 'CR' (Cheltenham Railway) began to appear in Brunel's appointment diaries, the project adding another burden to Brunel's already enormous workload and providing considerable engineering challenges as it crossed the Cotswolds.

The promotion of the Oxford branch line was less straightforward. GWR proprietors had been told in 1836 that a branch to Oxford, with a continuation of it to Worcester, was being 'promoted by the leading interests of those cities and the best exertions of the Company will be devoted in cooperation with them to accomplish these objects'.[3] Plans for building a railway as far as Worcester did not appear in a bill put before Parliament in 1837, which instead proposed a railway from Didcot that would run to Oxford, which also included a short branch to the nearby market town of Abingdon. Brunel was able to pacify initial opposition to the scheme by revising the original site of the terminus and by providing assurances to the University, which, like Eton College, was nervous about the effect of the new railway on the morals of its students. This smoothed the passage of the bill through the House of Commons, but the opposition of two prominent figures who owned almost half the land needed to build the line ensured that it was rejected by the Lords.

Another attempt to get the bill through Parliament was undertaken the following year, this time with the Abingdon branch omitted. To support the bill, Brunel produced fourteen pages of closely written 'proof' text recording the route and intended operation of the railway, which was subsequently submitted as evidence to House of Lords.[4] There were no particular engineering difficulties in the proposed line he wrote, and gradients were 'generally favourable' with the steepest gradient being a gentle 1 in 660. In a supplementary note, he continued: 'The works of the Great Western Railway are proceeding with great activity.' He believed that the line from London 'would be complete as far as Didcot before the Oxford Line can be made'.

The 1838 evidence also contained petitions for and against the railway that provide further proof of the difficulties faced by Brunel and the GWR in those early days. Mr Thomas Sheard, described as a 'very respectable grocer' from Oxford whose chief business was with the University and private schools, was in favour of the railway. He reported that the bulk of his supplies then came from London, mostly by horse-drawn wagon which took at least two days; when goods were moved by canal or river though, the process took about a week allowing for delays at the wharf. There were further submissions by other traders and farmers, all supporting the railway on the basis that it would improve their business prospects considerably; William Hopkins, a coal dealer and 'wharfinger' from Oxford, gave evidence of the inconvenience which frequently arose from 'delays on the rivers and canals from flood and frost', adding that the new railway would be of very great importance to the City of Oxford.[5]

University authorities were less supportive and more parochial in tone. A petition from the 'Chancellor, Masters and Scholars of the University of Oxford' noted that they felt satisfied that 'the means of communication already in existence between Oxford and London are fully adequate for the purposes of trade and the conveyance of passengers.' Any more access to the metropolis than currently enjoyed by students would be 'extremely injurious' to the discipline of the University they continued, adding that the proposed railway had not originated from the 'wants and wishes' of the inhabitants of Oxford and was being promoted by 'strangers unconnected with its interests...in a spirit of speculation'.[6] Similar sentiments were shared in another petition from 'Inhabitants & Householders within the City of Oxford and its immediate neighbourhood'. The railway was generally unpopular with the citizens of Oxford and the University, the petition argued; there was a perfectly adequate turnpike road to Steventon where those who wished to use the railway to travel to London could join a train if they wished. The railway would lead to the destruction of comfort of many of the inhabitants of the city and eventually prove injurious to the health of the place. As something of an afterthought, the petition also mentioned that the construction of the line so close to the city would put 'humbler classes' of citizens in danger from flooding.[7]

While the snobbery and insularity demonstrated by these petitions was not shared by all Oxford residents, it was nevertheless true that Brunel's branch line to Oxford was largely to be financed by shareholders with little or no connection to the city. Evidence submitted to Parliament in 1838 reveals that £100,000 of the £120,000 projected cost of the new line was to be subscribed by ten people, all investing £10,000 initially. There were, the parliamentary submission recorded 'no shareholders who may be considered as having a local interest in the line'. Six of the investors were familiar names from Bristol and Directors of the GWR: Thomas Guppy, William Jacques, George Jones, James Lean, Peter Maze and William Tothill. The other four shareholders were Frederick Barlow, Robert Gower, Frederick Ricketts and Henry Simonds, who were from either London or Reading.

The bill was defeated again, with the opposition of the University and its formidable Chancellor the Duke of Wellington proving too much. After a further try in 1840, Brunel and the GWR abandoned the Oxford line for the moment, concentrating on the more urgent task of completing the London to Bristol line itself. For the time being, residents and students from Oxford's dreaming spires had to be content with a bumpy carriage journey from the city to Steventon to catch a GWR service that took ninety minutes and cost three shillings. By 1842 the University of Oxford had changed its position, and with only two of the colleges still objecting to the idea of a railway, a bill was deposited in Parliament for the 1843 session. This time the Act had an easier passage, and despite predicable resistance from canal companies it was passed on 11 April 1843. The relatively easy nature of the countryside, lack of major engineering works, and an unseasonably mild winter meant work proceeded quickly and the line opened on 12 June 1844. At just under ten miles long, the branch ran not to Steventon, but to a new station further west at Didcot. Initially at least, Oxford had a very modest wooden station which was not replaced with something more in keeping with the city until 1852.[8]

For a few short years while Oxford remained cut off from the Great Western Railway main line, Steventon became a hive of activity, as carriages from the city picked up and deposited passengers bound for the capital and the West Country. It seems likely that Brunel had already planned that trains bound for Oxford would eventually join the main line at Didcot rather than Steventon, as the station built at the latter location, despite being described as a 'first class' by the company, was also a timber structure that was relatively cheap to build and capable of being relocated elsewhere if necessary. The station and the line from Reading were inspected by the Directors on 11 May 1840, and trains began running from yet another temporary terminus of the railway at Steventon at the beginning of June. Eight trains a day were initially run, connecting with coaches operated by an Oxford carrier, Costar & Waddell.[9] The GWR did little to actively encourage this traffic but it suited them to ensure there was a connecting link with the city; in a short few years, though, there would be little need to worry as the completion of the line from Bristol to London would wipe out most of the coach trade entirely.

Brunel designed a handsome house for the Superintendent of the Line at Steventon, complete with ornate bay windows similar to those used at stations such as Bath. At just over fifty-seven miles from Paddington, Steventon was a halfway point on the railway, and after its full opening in 1841 and the subsequent abolition of separate Bristol and London committees the engineer was asked to make alterations to the house to enable it to accommodate board meetings.[10] This was duly done, but Steventon's role at the centre of all things GWR was to be short-lived and meetings were only held there for six months, after which power decisively shifted to Paddington where it would remain until the company was absorbed into British Railways in 1948.

The success of the Cheltenham bill would no doubt have helped to support the confidence of the GWR Directors and in the spring of 1838 most of the remaining contracts had been let for the work between Reading and Didcot, with Grissell & Peto undertaking much of the main work with the bridges at Basildon and Moulsford being built by Chadwick. Contracts 3R and 4R, consisting of earthworks between Moulsford bridge and Didcot, were let to Custance & Bedborough, while the construction of stations between Steventon and Corsham was awarded to the London contractor J D. & C. Rigby.[11] As contractors continued to construct the new line, Brunel and the Company continued to struggle with the demands of greedy landowners keen to maximise compensation for both the short-term inconvenience they might suffer while construction work took place and the perceived loss of value they might incur from the railway in the long term. A surviving exchange of correspondence between a landed client and an assessor or solicitor provides an interesting example of we might now call compensation culture. Writing in June 1839, Mr W. Taylor of Didcot highlighted the amount of land affected by the construction of a cutting which split a field in two. Compensation of more than £68 would be due, including an allowance for trees cut down by contractors as well as loss of access to land while the work continued. But the letter concluded, 'you had better ask them more and get as much as you possibly can'; it was no wonder that the amount originally set aside by Brunel for land purchase and compensation proved woefully inadequate and contributed to the increased cost of the railway.[12]

A month after the line had reached Steventon the line was completed to Faringdon Road (later renamed Challow) near Swindon.[13] From there it was a short journey to Swindon, which before the coming of the GWR was described as a small market town 'set upon the summit of a moderate hill' at the western end of the Vale of the White Horse. In later years the Great Western Railway would describe old Swindon as 'an ancient market town of some note', and although it was true that it had been mentioned in Domesday Book, with a population of around 2,500 and being 'adorned with the mansions of several persons of independent fortune' it had been losing influence to other local towns like Highworth and Wootton Bassett before the arrival of the GWR. By the end of 1840, there was no station in Swindon itself and the construction of Brunel's broad gauge railway had continued through the fields north of the town and had reached a temporary terminus four miles west of Swindon at a place called Hay Lane. The new station named Wootton Bassett Road by the GWR was close to the turnpike road running between Swindon and Wootton Bassett. Brunel told shareholders in August 1840 that the new temporary terminus should be complete by November, but it finally opened without ceremony on 17 December.

With the line between Bristol and Bath now complete, the report of the tenth half-yearly meeting of proprietors concluded that the opening of the railway to Hay Lane would leave only thirty miles of road to be travelled by carriages, bringing the completion of the line ever closer. Contracts were again agreed by the GWR with coach operators in Bath and Bristol to run horse-drawn services until the line was complete, but similar arrangements to operate coaches from Hay Lane to Cirencester and Cheltenham were not proceeded with due to the poor state of the turnpike roads north of Swindon.

Today there is no evidence of what must have been a substantial station at Hay Lane; when the Board of Trade Inspector Sir Frederic Smith visited in early December 1840, he noted that 'although Hay Lane station is merely intended as a temporary terminus, the company are forming it, in regard to sidings, switches and other mechanical arrangements, in the same extensive and substantial manner as in their ordinary practice at permanent terminals.'[14] It seems that Smith was referring to a complex that was rather more than a station, and while there is little archival record of Hay Line, it appears that the station was provided with refreshment rooms and stabling for horses, as well as locomotive facilities for 'iron horses' including a substantial running shed or 'engine house'.[15] Brunel sketched out layouts for an engine house[16] at 'W B Road' in early 1840 and also designed a row of twelve timber-built single-storey cottages that were erected at Hay Lane by the contractor Gandell. These temporary structures for the use of staff were not moved into Swindon when the new workshops opened but were relocated to nearby Eastcott Lane and were only demolished in 1939.

Brunel and Gooch had for some while been considering locations for a locomotive works. Following his appointment in August 1837, Daniel Gooch had spent several years setting up sheds and repair shops at both West Drayton and Paddington, as well as doing his best to maintain the fleet of workable and reliable engines from the collection of freaks and misfits already ordered by Brunel, so was well aware of the need for more locomotive facilities. With the opening of the London to Bristol line now approaching, and with agreements for the GWR to operate both the Cheltenham & Great Western Union Railway and the Bristol to Exeter Railway, the urgent need to have a central workshop to maintain and ultimately build locomotives was becoming pressing. Brunel had considered Reading and Didcot as potential locations but seems to have accepted that any new facility needed to be nearer a place where engines could be changed in order to tackle the steeper gradients faced by the railway at the west end of the line.

As late as June 1839, Swindon had apparently still not been considered as a location for a railway works, but – perhaps more circumspect following the difficulties resulting from his original locomotive specifications – Brunel asked the ever practical Gooch to report on the best place for a railway works; on 13 September 1840 Gooch wrote to Brunel reporting in favour of Swindon as the location for the Great Western's 'principal engine establishment'. He argued that Swindon was an ideal place for a works since it would be at 'a convenient division' of the GWR line where engines could be changed for the steeply graded section west of the town towards Wootton Bassett, Chippenham and Box, and it was already a junction for the proposed Swindon to Cheltenham line. He rejected Reading as too far away. Gooch told Brunel that coal and coke could be delivered to the works from the nearby Wilts & Berks canal 'at a moderate price', and that the canal could also be a potential source of water, although at that time he cautioned that he was 'not sufficiently acquainted with the place to know how far we might be affected by the want of water'.[17] These reasons, he concluded, made Swindon 'by far the best place we have for a Central Engine Station'.[18] Gooch also included a rough sketch showing the location and layout of a works situated in a 'V' between the GWR and Cheltenham lines.

Gooch and Brunel travelled together to Swindon shortly afterwards to visit the site of the proposed engine establishment, described by Gooch in his diary then as 'only green fields' and Brunel agreed with him that this was indeed the place.[19] It is probable that they ate their lunch in the 'furze, rushes and rowan' in what was then still open country, but the oft-repeated myth that Brunel threw down a sandwich and claimed this as the place where the works should be built is just that, a myth. Brunel trusting his engineer, reported his conclusions to the Board of Directors who formally approved Swindon as the location for its Principal Locomotive Station & Repairing Shops on 8 October 1840. Shareholders learned of the decision at a General Meeting in February 1841 when they were told that the Directors would be providing 'an Engine Establishment at Swindon commensurate with the wants of the Company, where a change of engines may advantageously be made'.[20]

Brunel and Gooch were then faced with the task of designing, building and equipping not only a large engineering workshop complex to be used for both maintenance and heavy repair of locomotives, but also the more urgent task of providing a running shed for engines hauling trains on both the main line and the Cheltenham branch. Although Gooch had originally sketched out a 'roundhouse' shed design in his 1840 letter, Brunel's original designs for engine facilities had featured a rectangular structure of around 500 ft in length with workshop blocks at either end.[21] Evidence from various sources now suggests that the large engine shed originally erected at Hay Lane may well have then been dismantled and moved to Swindon Works as a temporary measure.[22] *Herepath's Railway Journal* for November 1841 reported that 'the whole of the Locomotive Department of the GWR has been removed from Wootton Bassett to Swindon' and a later list of railway staff and contractors covering the period from 1843 to 1865, compiled by John Fawcett, recorded Henry Appleby as 'first Superintendent of the Running Department' and that 'the "A" Shed was removed from Hay Lane before the new works were completed.'[23]

The engine shed was then rebuilt and gradually extended between 1841 and 1846 and once completed was able to house at least forty-eight locomotives and tenders; it remained in use until 1871 when a new more modern shed was built to the east. The old Brunel building survived as a workshop until 1929 when it was demolished but surviving photographs show a timber panel construction that could easily have been transportable from Hay Lane to Swindon in the 1840s. An account of the works published in 1852 described the engine house as a place where locomotives were kept, 'like horses at livery, except they require no food', a building resembling 'a veterinary college where any constitutional defects could be corrected and any local injuries repaired'.[24] The first section of the railway works took a further year to complete and at its core were a series of workshop buildings grouped around three sides of an open courtyard. Heavy machinery was supplied by the Whitworth Company of Manchester and was in operation by the end of November 1842, although the works was not formally opened until 1 January 1843. Despite financial difficulties suffered by the GWR in the aftermath of its opening, Swindon continued to grow, as the new network expanded and traffic increased. New buildings housing blacksmiths, machine and turning shops were added, along with an entirely new courtyard north of the main complex with new smithies, the workshops being equipped with two large Nasmyth steam hammers for forging locomotive crank axles. Finally, another range of workshops was added to the west side of the works accommodating facilities for wagon repair and construction. The new works initially had a staff complement of 423 that reflected its dual role as both engine shed and workshops; as a result there were ninety-eight footplate staff and sixty-six cleaners as well as general labourers, fitters, turners and erectors, coppersmiths, blacksmiths, engine painters and boilermakers.

Brunel was closely involved with the design of the works at Swindon, and his sketchbooks contain various ideas and designs for buildings and equipment; his letterbooks contain correspondence with his staff and external contractors and manufacturers of specialised equipment to be used in the factory. Still very busy supervising the completion of the railway and its operation following opening, Brunel appointed R. J. Ward as his resident engineer in August 1841, and the faithful Hammond also seems to have assisted in supervising work at Swindon on occasions. Gooch, with his experience of working at the Stephenson locomotive works at Newcastle and at other factories, would have provided invaluable help in both the design and construction stages and was instrumental in ensuring the new facility was equipped with the most up-to-date machinery and Whitworth tooling and gauges.

The perilous financial position of the Great Western Railway in the period after 1841 probably influenced Brunel's designs for the workshops at Swindon. There was simply not the money to create anything with any architectural pretension and so most of the buildings featured a standardised construction method that consisted of stone piers supporting timber panels into which doors or windows could be set where required. This 'pier and panel' system was economical to build and maintain. Foundations of many of the workshops sat on clay, and so the piers were supported by inverted brick arches at ground level; surviving examples of these can be seen on the front elevation of what is now the STEAM Museum, a building originally constructed in 1846. A standardised design was also provided for roofing, a 45 ft span queen post truss roof topped with Welsh slate being the norm. Most of the works was built of limestone brought to the site from the quarries of Pictor & Brewer in Bath and Corsham, along with sandstone from quarries in Swindon itself.

When the GWR Directors had authorised the construction of the engine establishment and repair shops in February 1841 they also announced that work at Swindon would include a new station and refreshment rooms as well as 'cottages etc. for the residence of many persons employed in the service of the company'. The financial difficulties being experienced by the company as a result of the ever-growing cost of Brunel's railway meant that the Directors also felt the need to reassure shareholders that the company would pay only for the construction of the railway workshops, the station, refreshment rooms and houses for the workforce would be built at the expense of London contractors J. D. & C. Rigby, who were already engaged in building the works. Company records note that two other London contractors, Chadwick & Bailey and Shaw & Smith, had also been asked to tender for the contract to build the station and cottages, but Rigbys were probably chosen because of the important work they had already done in constructing Brunel's new station at Slough, and intermediate stations between Steventon and there.

While the reality was that the GWR were in no position to pay for the construction of the station and railway cottages, the two agreements ultimately signed by the company with Rigbys would be a source of inconvenience and irritation to them for almost fifty years. The initial contract of 2 February 1841 meant that at their own expense, Rigbys would build Swindon Station and refreshment rooms along with an estate of 300 houses to accommodate workers from the new Locomotive Establishment. The deal seemed to have appeal for both sides. The agreement stated that Rigbys would be entitled to the profits generated from the operation of the refreshment rooms and hotel facilities to be provided, and that the GWR would also lease the cottages from them and pay rent on a quarterly basis. In return, Brunel and the Great Western Railway would have a new and large station complete with grand refreshment rooms at no direct cost, along with housing for its Swindon workforce.

Although the £15,000 cost of the station was borne by Rigbys, it was Brunel who was responsible for its design and specification. From early sketches, he produced a series of plans that featured a station consisting of two separate buildings linked by a walkway at first floor level, mischievously likened by Rolt to the 'Bridge of

Sighs', that allowed passengers to cross the tracks to change to and from Cheltenham trains and gain access to hotel facilities that were initially provided on the top floor level.[25] The simple but elegant, classically styled buildings were constructed from local Swindon sandstone, and initially covered in render. As the line was carried on an embankment at this point, passengers ascended stairs to the platforms, while underneath, kitchens, wine cellars and staff accommodation was provided at basement level. The station was not complete by the time the GWR opened from Paddington to Bristol in May 1841 although temporary facilities appear to have been provided before the official opening of the station by the Duke of Cambridge on 14 July 1842.

Having little experience designing refreshment rooms and grand interior décor, Brunel had approached Francis Thompson, engineer and architect of the North Midland Railway for advice, since he is thought to have worked on the design of refreshment rooms at both Wolverton and Derby. Writing in January 1842, Brunel asked him to send an invoice 'for the trouble and expenses incurred by you on the subject of refreshment rooms'. The engineer's usual need to impress seems to have been much in evidence, if early illustrations of the Swindon Refreshment Rooms are to be believed. The first-class rooms were particularly grand, Brunel's original specification noting that they should be 'handsomely fitted up' and decorated with imitation marble columns and pilasters, with an ornate chimney piece topped by a mirror. Writing later in the 1840s, S. C. Bree reported that the facilities were 'magnificent' and that the tables were 'always spread with different viands and refreshments'. Not surprisingly, accommodation for second-class passengers was somewhat more modest; the walls were papered rather than ornately decorated, woodwork was of grained oak, and the floor was linoleum rather than carpet.

Despite the grandness of the surroundings, the standard of fare served to both first and second class travellers was less than satisfactory. A second lease signed on 18 December 1841 had noted that the GWR should 'give every facility to Rigby's (sic) for enabling them to obtain adequate return' adding that all trains carrying passengers 'apart from emergency or unusual delay' would 'stop there for refreshment for a reasonable period of about ten minutes' and there would also be no other refreshment rooms on the line apart from at Bristol and Paddington.[26] Four days after signing this agreement, Rigbys sublet the lease for an annual sum of £1,100 plus a £6,000 consideration to Samuel Griffiths, an hotel owner in Cheltenham, presumably in an attempt to recoup some of the considerable expense they were incurring from the building of Brunel's grand station design. With no control over the operation of the refreshment rooms, the GWR began to regret the arrangement as complaints about high prices and poor service and food grew. Brunel famously wrote to Griffiths himself: 'I did not believe that you had any such thing as coffee in the place; I certainly never tasted any.' He signed off by telling the hotel owner that he had 'long since ceased to make complaints at Swindon. I avoid taking anything there when I can help it.'[27] Although the quality of food served did improve, the lease was sublet again on a number of occasions, and despite the best efforts of Great Western Railway solicitors, the company was unable to escape its obligations and the ten-minute stop became an embarrassment as improvements to train speeds and timetables took place in the course of the next fifty or so years. The difficulty caused by the stop was only ended when the lease was bought out by the GWR for £100,000 in 1895, more than thirty years after Brunel's death.

Rigbys had also been tasked with the construction of 300 cottages to house workers from the new railway factory and their families. Accommodation was badly needed as Swindon and North Wiltshire had little or no tradition of heavy engineering. Wyld's Guide of 1838 said of the town that 'no particular trade' was carried out there, but there were 'some extensive quarries which together with agriculture afford sufficient employment for the inhabitants'.[28] This meant there was no local pool of labour to draw on with any experience of either building or repairing steam engines or making or assembling components for them. Further west in Bristol there was an established shipbuilding trade, and Bath, despite its reputation as a spa, also had a number of factories and foundries; but Daniel Gooch, and his assistant Archibald Sturrock realised that more skilled artisans would be needed to ensure that the works at Swindon was successful. Gooch had already managed to tempt men from his native North East to work at his depots at Paddington and West Drayton, and further recruitment took place from there and in other areas where railways and engineering were already well established such as Liverpool, Manchester, Leeds, Scotland and South Wales.[29] While obtaining the services of such men was vital, retaining them was equally important, and so the provision of houses was an urgent priority.

According to the February 1841 contract J. D. & C. Rigby were to provide the 300 houses at their own expense on land provided by the GWR by Christmas Day 1842, to coincide with the opening of the railway workshops on 1 January 1843. In the event, this proved hopelessly optimistic and by December that year, Rigby's had only completed 130 cottages, at the western end of what is now known as the Railway Village. Both the railway and the contractor could share some of the blame for the delays; as had been the case on other contracts, Brunel adopted his normal penny-pinching attitude with contractors and he and the Company were slow to settle bills, especially as the costs of his railway continued to increase. Rigbys had also underestimated the difficulties they might encounter in contracting to build the railway works, station and railway cottages simultaneously, and struggled not only with cashflow, but also with finding enough skilled labour to complete all the work on time.

Despite an offer of assistance from Francis Thompson, Brunel had undertaken the design of the first railway houses in Swindon himself, the plans once again being drawn up by draughtsmen in his Duke Street office. The Directors approved his ideas for a first block of eighty-eight houses in March 1842 with each cottage not to cost more than £100. The energetic Brunel staked out the location of the first two rows himself and Rigbys set to work shortly afterwards. As designed, the railway houses were imposing structures that included Jacobean and Elizabethan motifs that featured on many of the engineer's larger buildings. Although in writing to Thompson Brunel he had described these first houses as having 'a totally unornamental character'[30] they were solidly built and a great improvement on anything similar provided by railways elsewhere. The relatively modest budget of £100 per house meant that the interiors of the cottages were very basic, however, with only one room on each floor, no kitchen and a privy in the small back yard. Although many in the town now believe that the cottages were built from stone brought from workings at Box Tunnel, it now seems likely that although some limestone from there was used for decorative features, most of the building material for the railway village came from Swindon quarries transported down the hill from the old town by a temporary tramway.

The slow pace of progress by Rigbys' labourers drew a predictably irritable response from Brunel in July 1842, when he told them that he would not provide them with any further money 'while the works are proceeding so unsatisfactorily'. The contractor was forced to sub-contract work on the next batch of forty-four houses and to help speed construction, Brunel made the design of subsequent cottages plainer but retained the same dimensions. Work began on a third block of houses at the beginning of 1843, these being completed by the end of the year, but no more houses were constructed until 1845. By this time, overcrowding in the houses in what became known as 'New Swindon' to differentiate it from the older settlement up the hill, was becoming intolerable as the number of men employed in the new railway works grew to around 1,800. In the end, it would be fourteen years before the 300 houses specified in the original agreement with Rigbys were completed. Brunel retained overall responsibility for the project, but later drawings seem to have been produced by contractors, with the work supervised on the ground by R. J. Ward. The street names reflected stations on Brunel's new railway, with Bristol, Bath (now Bathampton), Exeter and Taunton streets forming the western side of the village, and Reading, Oxford, and London streets to the east. As his responsibilities took him further away from Swindon, Brunel had much less to do with his creation, and Daniel Gooch as Locomotive Superintendent and eventually member of Parliament for the nearby constituency of Cricklade[31] played a more active role in its continuing development.

West of Swindon, Brunel and the hundreds of navvies toiling to complete the line had a far tougher landscape to cope with. Brunel had made enormous efforts to ensure the gradient between London and Swindon was as shallow as possible, managing to maintain a ruling gradient of 1 in 1320 leading to the stretch of line between those two places being nicknamed 'Brunel's Billiard Table'. After Wootton Bassett the route to Chippenham and Bath included a number of steeper gradients, deep cuttings and embankments and probably the most difficult work on the whole railway, Box Tunnel. The construction of the GWR had been badly affected by unusually wet winters and the winter of 1839-1840 was no exception, *The Railway Times* reporting that the rain had been 'incessant'. Brunel was compelled to report to shareholders in August 1840 that the severe weather had given rise to reports in newspapers of 'the impossibility of completing the line in the reported period'.[32] In particular he identified the heavy clay embankments in the Wootton Bassett area, cuttings and embankments beyond Chippenham, the Box Tunnel, earthworks in the Box valley, and other work needed to complete the railway's entrance into the city of Bath as having been particularly affected by what he called the 'wet season'. Work to form the deepest cuttings and highest embankments had been postponed until the weather was dryer, but even with this precaution, geological conditions had meant soil had continued to slip in and around Wootton Bassett and Brunel had been forced to ask for more funds to underpin earthworks with piling, an expensive and time-consuming process.

After Chippenham station, the railway ran along what Brunel described as an 'extremely long and very lofty' embankment for almost two miles,[33] and then continued through a deep cutting for a further three miles before reaching both his greatest challenge and greatest achievement on the GWR, Box Tunnel. Brunel's well executed plans had ensured that Box was the first tunnel to be encountered by travellers on their 96-mile journey from Paddington. At almost two miles long, it was the longest on the line, and its construction presented the most difficulties to the contractors and navvies who worked on the GWR. Less than a year after the passing of the 1835 Act, Brunel wrote a report for Directors telling them that 'the Box Tunnel is the most important work' and that it would determine the completion of the whole line.[34] Considerable discussion and debate about the tunnel, then thought very remarkable for its size of bore and length, had dominated many hours of the committee stages of the Bill and critics of both Brunel and the GWR were eager to pounce on any difficulties that the project might encounter.

Aware of potential trouble ahead and drawing on the experience he had gained working on the Thames Tunnel, he ordered the sinking of trial shafts to confirm the geology of Box Hill. In early 1836 seven vertical shafts between 70 feet and 300 feet deep were dug down to what would be rail level, enabling Brunel and his assistants to plot the location of the Oolitic limestone, clay and Fullers Earth that lay along the line of the tunnel. The contract to undertake this work was awarded to Paxton & Orton, whose price of £14,000; while £3,000 more than Brunel's estimate, it was still a modest sum considering the work required. An additional shaft was required at the east end of the tunnel and the contract for this was given to Lewis & Brewer, a local quarrying concern. Paxton & Orton, described by Brunel himself as 'not men with capital' struggled to complete the work. It is recorded that aware perhaps of the reputational and financial risk to both the GWR and himself, Brunel even resorted to paying Paxton & Orton out of his own pocket to support them. Even this desperate measure failed, and the contractors failed in April 1837.

These difficulties were compounded by reluctance on the part of other railway contractors to tender for the task of excavating Brunel's tunnel. His reputation for treating contractors poorly had preceded him and this, added to the obvious risk of providing a fixed price for a project fraught with unknown challenges, meant that many of the best-known firms did not compete for the contract. As a result, Brunel, the Directors and company shareholders had to endure a further six-month delay on what was already a challenging construction programme. Contracts to build the tunnel were finally let in February 1838, with the majority of the excavation being handled by the Kent contractor George Burge, who was to be responsible for all but 2,820 feet of tunnel, the remaining work being carried out by Lewis & Brewer. The Company Chairman William Sims put a brave face on what was clearly not an ideal situation in a half-yearly meeting held the same month, noting that the contracts for the entire tunnel were to be undertaken 'by contractors of character, ability, and property, and on terms satisfactory to the Directors and Engineer, although higher than had been originally contemplated'. He also noted that the work would be completed in thirty months, enabling the line to be open by the summer of 1840, but this ambitious target, like many proposed on Brunel's railway, was to prove wildly inaccurate.

Miners struck out on tunnelling faces from each of the shafts, which were equipped with horse-drawn cranes used to raise and lower men, materials and spoil from inside the hill. As a writer noted in *The Railway Times* in 1840, the difficulties encountered in the completion of the tunnel were 'appalling', the geology providing different challenges for both contractors. At the east end, the hard limestone through which the tunnel was driven proved to be a formidable obstacle; it was reported that Lewis & Brewer were using more than a ton of gunpowder every week to ten days to blast through the rock and while precautions were taken, accidents were frequent. There is no definitive record of the total number of fatalities but it is possible that more than one hundred workmen lost their lives during the six-year construction period. Local newspapers like *The Bath Chronicle* featured regular reports of serious or fatal accidents throughout the construction.[35] Brunel, appearing before a parliamentary committee in 1846 told MPs that a list of more than 130 workmen admitted to the hospital in Bath between 1839 and 1841 as a result of injuries sustained at Box was 'a small list considering the heavy works, and the immense amount of powder used'.[36]

Conditions for navvies were absolutely miserable; in addition to the difficulties encountered with limestone, they also had to contend with bands of Fullers Earth and blue Marl, thick, sticky clay which was difficult to work with. Temporary roof supports were required while miners dug out spoil with bricklayers following behind building a lining. More than thirty million bricks were eventually needed inside the great tunnel, most manufactured at the Chippenham end of the tunnel and delivered by a team of more than one hundred horses.

Some idea of just how unpleasant Brunel's tunnel was can be found in a description provided by an anonymous correspondent from the *Wiltshire Independent* printed in the *Railway Times* in 1840. He was lowered down one of the shafts into the tunnel where he found the atmosphere 'oppressive, unpleasant and stuffy due to the lack of fresh air and the smell of smoke and gunpowder'. The 'dark dim vault' he recorded, 'was saved from utter and black darkness by the feeble light of candles which are stuck on the sides of the excavation and placed on trucks and other things used to carry out the works'. He picked his way across a hazardous scene littered with temporary rails, stone blocks and pools of water accompanied by the sound of picks, shovels and hammers and was relieved to emerge from the gloom into the fresh air 'full of wonder at the skill, enterprise and industry' of the workmen.[37]

Water from underground streams and fissures made flooding a constant hazard more generally, *The Great Western Magazine* later reporting that the influx of water 'occasioned great expenditure of labour as well as considerable annoyance'. Extra pumps were required and Burge fell behind schedule in the autumn of 1838, incurring the wrath of Brunel. In a series of increasingly impatient letters Burge was taken to task by the engineer. On 4 September Brunel told the contractor that he was not sure if Burge had 'deceived him' as to his intentions of completing the work within the allotted time, or whether the delays were instead a result of apathy or negligence. Having visited the tunnel, he continued, the progress made appeared to him to be ridiculous. By January 1839 matters had not improved enough for Brunel's liking and he told Burge that unless he could 'report satisfactorily...upon general progress during the next fortnight no further confidence will be place in you'. Brunel had resorted to his usual tactic of withholding payment and enforcing penalties for falling behind with the work. In April Burge was told by Brunel that he would not recommend the Directors made any further payments and that he also disagreed with the accounts supplied by the contractor, noting that William Glennie, his resident engineer, would provide him with a corrected statement. Glennie, appointed by Brunel in 1838, spent many long hours in the dark recesses of the tunnel and was injured in an accident there in 1839.

Brunel was in no position to dismiss Burge, despite his frustration at the lack of progress. The contractor and his men continued to toil in intolerable conditions, but by August 1840 Brunel could report that five-sixths of the tunnel was completed. Excavations were not finished until the end of March 1841 and then only after additional workmen and materials had been drafted in to make the final push. It was reported that almost 4,000 men and 300 horses were crammed into the workings with Brunel personally supervising the last stages of the work from a base at *The Queens Head Inn* in Box. With the Bristol-Bath section of line already complete, the opening of Box Tunnel on 30 June 1841 meant that the Great Western Railway main line from London to Bristol was finally complete, although *The Railway Times* reported that passage through the tunnel was made 'with a degree of slowness and caution' since only a single track had been laid, and Gooch remained there for some weeks supervising the safe passage of trains.

Emerging from the grand west portal of Box Tunnel, trains then quickly dived back into a 200-yard tunnel at Middle Hill, which also featured a stylish classical design. The line then snaked down the Box valley to Bath. Before reaching the station there, Brunel had to engineer the passage of the railway through Sydney Gardens, a 16-acre park laid out in the eighteenth century, a fashionable destination for both local people and visitors to the Georgian spa and described as the city's 'Vauxhall' by a contemporary railway guide. The Kennet & Avon Canal had bisected the gardens in 1810, but its impact on the landscape was less dramatic than the intended route of Brunel's new railway line. It might have been expected that the owners of the gardens would have resisted the progress of the railway, particularly since its route would involve the demolition of buildings including the tea house and would block a number of paths across the parkland. Brunel was at his most diplomatic reassuring the Trustees of the gardens that the railway would actually enhance the park through his elegant architecture; although in the event, as Andrew Swift records, the proprietors of Sydney Gardens were short of money and the compensation they would obtain would be most welcome, as had been the 2,000 guineas they had received from the canal company some years earlier.[38] Rather than hide the railway in a 'covered way' as originally proposed, Brunel made it a feature, separating the line from the park with an elegant colonnaded Bath stone wall, and building two bridges, one in iron, the other in stone, to cross the tracks. Writing to one of the Garden Trustees in January 1839, Brunel also noted that the upper part of the park would feature a retaining wall; this massive structure was required to support the canal, situated close by.[39] In the process of these delicate negotiations, Brunel had also persuaded the Kennet & Avon Canal to divert the course of their waterway, to reduce the number of times the railway would cross it, which was an added bonus.

Although his designs for the station at Bath were not in the classical style, they were nevertheless intended to enhance the railway's importance to the city and provide an elegant and imposing entrance for passengers. Brunel's Jacobean style station was described by J. C. Bourne as having 'debased Gothic windows and Romanesque ornaments' but noted that its principal feature was its roof, which was 60 ft in length 'without buttress or tie of any description'. The overall roof covering was similar to the one provided by Brunel at Temple Meads but was removed in the late nineteenth century and few pictures survive of its original appearance. The station was opened when the line to Bristol was completed on 31 August 1840 and its brief life as a terminus ended the following year when the completion of the Box Tunnel allowed the whole railway to be opened.

In the interim Brunel had continued to busy himself with minor detailed designs one might have expected to have been done by junior members of his staff; sketches for items such as elegant lamp posts for Bath station were then drawn up by assistants. He also continued to have a troubled relationship with some of his assistants and resident engineers, especially those he considered to be lazy or inefficient. One such unfortunate was Fripp, who infuriated Brunel on a number of occasions, resulting in memorable correspondence that has been much quoted. Fripp had been working on the Bristol to Bath section of line, and in October 1840 Brunel had already warned him in no uncertain terms about his attitude and lack of application. Fripp had clearly not changed his ways eight months later when his boss wrote again to complain about instructions he had given over Bath Station. 'I am sick of employing you to do anything that, if I had ten minutes to spare, I would do myself.' He instructed the indolent assistant to go to Bath to sort the matter, concluding that if he did not, 'pray keep out of my way or I will certainly do you a mischief.'

Leaving Bath, the railway then passed over a striking castellated Gothic viaduct beginning what one contemporary writer described as 'one of the most interesting lines in the kingdom from an engineering point of view'.[40] In the course of eleven miles of railway, Brunel faced a wide range of challenges that included four tunnels, numerous deep cuttings, five river bridges and more than thirty road and occupation bridges that crossed over and under the railway. In 1870 his son had noted that 'the nature of the building stone' had enabled his father 'at a moderate cost, to make the bridges, tunnel fronts and station ornamental features in the picturesque scenery through which the railway passes'.[41] The line was restricted by the narrow valley of the River Avon and at Foxes Wood even required a short diversion of the waterway itself.

The line between Bath and Bristol contained a number of striking and impressive tunnels. At Twerton there were two in quick succession; the first, only forty-five yards long, was almost a 'covered way' but it was followed by the 264-yard Twerton 'Long' Tunnel. Both east and west portals featured a gothic arch set into substantial structures with flanking towers redolent of Tudor fortresses. Shortly after the tunnel, the line ran into a 30-ft-deep cutting where in January 1838 the remains of a Roman villa were found as navvies dug the earthworks. One might have expected that archaeological material of this type might have been seen as an inconvenience to the progress of the railway, but both Brunel and the Directors were sympathetic to the preservation of ancient material found during excavations, and he instructed one of his staff in December 1837 'not to give away our Roman remains' as it was intended to set up a museum to display such finds.[42] It was suggested that the remains found there might have been that of a villa owned by the Roman governor of Bath. Excavations revealed metalwork, weapons, knives, vases, pottery and glass. Before they were swept away, plans of the Roman buildings were made and two well-preserved tessellated mosaic floors were removed and reinstalled at Keynsham station a few miles down the line. J. C. Bourne was not impressed by them, however, describing the mosaics as 'very rude, both in design and construction'.[43] There they remained until 1851 when they were donated to Bristol Museum.

The railway continued through another tunnel at Foxes Wood, known as No.3 tunnel, which at 1,017 yards long was the longest on the line apart from Box, the railway cutting through the Pennant sandstone on a long curve. Bourne's engraving of the west portal shows a picturesque scene with the River Avon close by and yet another castellated tunnel mouth. The No.2 tunnel situated at St Anne's was much shorter at 154 yards long, but at its west end had a portal that demonstrated Brunel's sense of theatre. The crenelated parapet of the tunnel mouth was damaged by a landslip during its construction and as his son recounted in his 1870 biography, 'The structure was left unfinished, and was planted with ivy so as to present the appearance of a ruined gateway.'[44] Bourne added that the tunnel mouth was serviceable and picturesque, although it appears that this piece of self-indulgence did not last long and the masonry was repaired after a few years.

After passing through what was described as a 'secluded dingle', Brunel's railway came at last to the outskirts of Bristol, making its entrance into the city by way of No. 1 tunnel, the last on the line. At the west end of the 330-yard-long tunnel was what Wyld called 'a noble circular arch enriched with a massive cable twist moulding, supported by plain shafts of fine proportions'. The tunnel mouth was on the scale of those at both Box and Middle Hill, and its Romanesque design was a contrast to the largely Gothic theme Brunel had chosen for most of the tunnels and bridges on the preceding section of line.[45] The east portal featured two other artefacts rescued by navvies when the tunnel was excavated. Two large stone nodules were found in the sandstone and Brunel and the Resident engineer George Frere arranged for them to be set on pedestals either side of the tunnel mouth. Nicknamed the apple and pear stones they remained until 1889 when the tunnel was opened out into a cutting; one still survives and is now displayed outside part of the University of Bristol. The railway continued through the red sandstone strata in a fifty-foot-high cutting, crossing the River Avon for the last time across a handsome three-span Gothic bridge before continuing to Brunel's terminus at Temple Meads.

Since the line from Bristol to Bath had opened the previous year, public celebration in the city where the Brunel's railway had been born was relatively muted on 30 June 1841 when the final section of line between Chippenham and Bath including Box Tunnel was opened, allowing trains to run from the city to London without interruption. After almost six years of struggle, Brunel's magnificent vision had finally become a reality, but the opening of the 112-mile line was marked with relatively little ceremony; a Directors special left Paddington at 8am that morning, arriving in Bristol four hours later. To have travelled such a distance so quickly was no mean achievement, but even by 1841 the novelty of railways was perhaps beginning to fade, for the inhabitants of Bristol and London at least. *The Railway Times* congratulated the Directors on the 'completion of this magnificent undertaking' but could not resist a jibe aimed at Brunel, concluding that they would for now 'sink all differences of opinion on points of construction and management'.[46]

6

Bristol and Beyond

Almost fifty years after Brunel's death, a Great Western Railway correspondent Alfred Arthurton wrote that 'the importance of Bristol as a railway centre can scarcely be exaggerated.'[1] Situated at 'the confluence of streams of traffic from all four corners of the kingdom' it was a key location on the network, but in 1841 it was merely the end of a journey on the original line from Paddington. While Brunel had envisaged the construction of a network of interconnected broad gauge railways across the country, the site he chose for his terminus at Temple Meads turned out to be too small and constricted for the growth of passenger and goods traffic that the arrival in the city of other railway lines he had engineered would eventually precipitate. It was also, as generations of frustrated and weary travellers will confirm, inconveniently situated outside the city walls and more than three-quarters of a mile away from the heart of Bristol.[2]

In hindsight it seems strange that even though Brunel was actively involved in the planning and construction of the Bristol & Exeter and later the Bristol & Gloucester railways at broadly the same time as he was working so hard to complete the GWR main line, as his office diaries record, he does not seem to have considered what implications these new enterprises might have for his Bristol terminus. Although it is true that the concept of a truly national railway network was still some years away, the fact remained that from 1835 not only did Brunel regard his 'finest work in England' as the first in a series of interlinked lines, but Great Western Railway Directors also clearly had wider commercial ambitions, investing in and providing support to other railways which could attract more passenger and goods traffic to their own line. By the beginning of 1841, with the opening of the GWR imminent, Brunel may well have begun to ponder the implications of his plan, writing to Locomotive Superintendent Daniel Gooch telling him of his 'considerable anxiety' of a time fast approaching when the railway would be operating as 'a concern of most unusual magnitude'.[3] By then however, it was too late to change plans for Bristol and the construction of his new terminus there was almost complete.

The site finally chosen at Temple Meads was in reality the nearest piece of open land remaining close to the city and pushing the railway further into Bristol would have involved complex negotiations, legal fees, and the payment of compensation to landowners. In 1835 Brunel had suggested that the new station could be situated at Old Market, closer to the city centre, but this would have entailed driving the railway through a heavily populated commercial and residential area which was judged not to be acceptable given the experience he and the Directors had already gained in planning the railway's entry into London. The land selected at Temple Meads was level and still largely open fields used for grazing, with some associated small industrial concerns. Brunel took three GWR Directors to the top of St Mary Redcliffe Church in 1836 to look at possible sites, and it was agreed that Temple Meads was the only realistic location for the passenger station, goods shed and other facilities. He later noted that the Directors had agreed with him that Temple Meads was the best they could select as a terminus for the railway.

Matters should have been made easier by the fact that most of the nineteen acres of meadow needed were owned by Bristol Corporation, but the process of acquiring the property was a protracted affair. In May 1836 GWR Directors agreed to go ahead with the purchase at a cost of £11,300, but this valuation was disputed by the Corporation and after some wrangling a final agreement for the transfer of land was agreed at a cost of £12,000 in December. The purchase had not, however, included the acquisition of a cholera burial ground, scavengers yard and a large part of the cattle market, not the kind of neighbours envisaged by Brunel for his prestigious Bristol station.[4]

The Temple Meads site was also bounded on two sides by the River Avon and the Floating Harbour. Speaking about his choice of site, Brunel reported that 'there was ample room for sheds and warehouses, that had a long extent of wharf...along which a great portion of the Kennet & Avon trade now passes' and noted that he had consulted the Directors of the railway, who, 'being commercial men connected with the trade of Bristol, must be well acquainted with the point that would suit the trade of Bristol best'.

With the location of his Bristol station agreed, Brunel began the work of producing plans and designs for its construction. As before, these started as sketches which were then worked up by draughtsmen at Duke Street; once completed, drawings and estimates were submitted to the Bristol Management committee for consideration, who appeared to have a more liberal view towards the sometimes extravagant work proposed by their engineer. The new terminus and the line towards Bath needed to be between fifteen and twenty feet above ground level to enable it to cross the river and the Floating Harbour, and so the station was carried on arches, with platforms at first floor level, making its arrangements 'somewhat peculiar' J. C. Bourne reported in 1846.[5] Brunel's proposals incorporated three distinct functions for his station; firstly a large covered train shed with platforms, secondly an engine shed, and thirdly an administrative block facing the road, which could also serve as the entrance for passengers.

As always, Brunel was anxious that the architectural design of his new station should reflect both the prestige of his new railway and its importance to the ancient city of Bristol, with costs a secondary consideration. In May 1839 he submitted two designs for the offices which were discussed by Bristol Directors some months later. Aware of the rising cost of Brunel's railway, and the more frugal reputation of their London counterparts, Thomas Osler, Secretary of the Bristol Committee, wrote a report on Brunel's plans to Charles Saunders. His committee was aware of the importance of 'economical construction' he reported, adding that they had considered two designs, the first in a Tudor style, and the second consisting of 'as thoroughly naked an assembly of walls and windows as could well be permitted to enclose any Union Poor House in the Country'. The Tudor option would match the character of bridges and tunnels already planned for the line from

Bath, he added, and would only cost £90 more than the 'Quaker' option.[6] Osler hoped that the London colleagues would agree with their decision he concluded, noting that copies of the plans would be sent to Paddington for consideration.[7]

George Henry Gibbs had recorded a conversation in his diary in May 1839 with one of the Bristol Directors, Robert Bright, about 'the importance of keeping down the cost of the Bristol depot, particularly of all ornamental work'[8] and so was probably not happy with Brunel's initial Gothic design which had a massive three-storey façade evocative of one of Henry VIII's Tudor palaces, complete with battlements and towers flanking either side of the building. When tenders for this grand edifice were received, Bristol Directors took fright as the cheapest quote was over £9,000, and Brunel was sent away to produce something more modest. While clearly disappointed, he did not respond in the same perfunctory way he had when his proposals for Reading station had been rejected earlier in the year, but instead produced a scaled-back design that retained the Tudor theme but provided less facilities than his original plan at a reduced cost of just under £7,000.

Brunel's Tudor revival façade still looked very different to the more austere neo-classical design of Stephenson's stations on the London & Birmingham such as Euston and Birmingham and attracted praise and criticism in equal measure. George Measom described the station as 'the most magnificent yet existing on the line' in his 1852 guide to the railway,[9] while J. C. Bourne argued that the station was a building of 'considerable pretension', and that the style of the architecture was 'with some modification, in vogue in England in the reign of Elizabeth'.[10] Much later, the architectural historian Nicholas Pevsner would describe the frontage as 'incongruous indeed for a railway terminus'.[11] Writing in 1843, only twelve years after helping Marc and Isambard with their Clifton bridge proposals, Pugin savaged his architecture, mocking his 'castellated work, huge tracings, ugly mouldings' and 'no meaning projections' that made up a design that was 'at once costly, and offensive'.[12] Brunel was not the only engineer singled out for criticism, however, as Pugin clearly had little time for railway architects generally, concluding that they were prone to 'showing off what they could do, instead of carrying out what was required'.

Separate contracts for the construction of the office block and train and engine shed appear to have been tendered for, although Binding[13] records that little archival evidence survives of the latter. A contract was let with Brown & Son for the office building in January 1840 with a stipulation that all work should be completed by September that year, but work appears to have proceeded slowly; in June, a GWR sub-committee's minutes recorded that lack of progress on the passenger shed had been 'anxiously considered' and the secretary was 'directed to write to the Chief Engineer calling his particular attention to this subject'.[14] In any event, the station was not complete by the time trains began running between Bristol and Bath on 31 August 1840, passengers using temporary platforms and in May the following year, with the full opening of the railway only weeks away, work was still behind schedule and Brunel was forced to inform Directors that 'arrangements in Temple Meads will not be completed by the 31st, but that every exertion has been made, and will continue to be made to expedite it'.[15] The station seems finally to have been ready for use when the Great Western Railway opened from London to Bristol on 30 June 1841.

Brunel's terminus was cleverly designed to make the maximum use of limited space. Passengers arrived at the station, their carriages passing through the left of two gateways situated either side of the main frontage and were deposited into Clock Tower yard[16] leading to waiting rooms, ticket offices and other facilities located in the arches underneath. There was no station concourse and so separate staircases for first- and second-class passengers took them to platform level and the trains, with luggage being hauled up by hoist. Cabs and carriages continued through a short tunnel under the station where they could then collect passengers who had just arrived, before leaving the complex through the right-hand gateway. The train shed situated at the north-east end of the terminus contained two separate platforms for arrivals and departures with five tracks between them; these were unobstructed by columns under Brunel's magnificent seventy-two-foot-wide timber roof, constructed in the style of Westminster Hall. The mock hammer beam roof was supported on either side by a cast iron colonnade and although spaced ten feet apart, the columns were less than four feet from the platform edge, making them an awkward obstruction for passengers when carriage doors were opened.[17]

Pevsner noted that 'the light comes in in a most un-medieval way by glazing along the ridge'[18] with three skylights providing illumination in what must have been a smoky environment, particularly as Brunel's 'comprehensive railway building' also included a small engine and carriage shed. After arrival, locomotives moved forward from their trains to take on water and coke, and ash from the firebox could be dropped through a chute to the floor below. The roof was lower at this point, and steam and smoke seeped through cracks in the celling making the space above, used as a drawing office, very unpleasant at times. With a façade of ashlar Bath Stone the three-storey office building housed offices, a board room and initially accommodation for the Station Superintendent.[19] The shields of the cities of Bristol and London which sat close to the roof parapet when the building was completed remain there today, although showing the wear and tear of almost two centuries of traffic; both the offices and train shed remain as 'one of England's great railway relics'.[20]

By the time the GWR opened fully on 31 May 1841 work on the next link in Brunel's plan to spread his system westward, the Bristol & Exeter, was already well advanced and within a fortnight trains were running from Bristol to Bridgwater, starting from an entirely separate station built at right angles to the GWR terminus. In retrospect it does seem strange that Brunel did not anticipate the need for a larger joint station in Bristol, particularly as plans for the Bristol & Exeter had been discussed only weeks after the Great Western Railway Bill had been passed in August 1835. The secretary of the provisional committee for the B &ER wrote to Brunel on 17 October inviting him to contribute 'engineering direction' to the planned railway and suggesting that an application to Parliament should be made as soon as possible.[21] Brunel was clearly anxious to oblige, and four days later the Bristol & Exeter company published a prospectus in which he was named as engineer of the new railway. It seems likely that Brunel may have already given some thought to the route of the line before discussing it with the committee, as the prospectus stated with characteristic Brunelian confidence that the 'line of this Rail Road presents no peculiar obstacles, and that a great part of it will be carried through flat country'. The advice of the engineer had been that the route should be formed through level country towards the

Bristol Channel as near to Clevedon and Weston as possible, and then on through Bridgwater to Cullompton and Exeter.[22]

The rapid pace at which the venture was moving continued and on 28 October 1835 it was agreed that an application to Parliament would be made in the next session, which gave Brunel only a month to select and survey a route and complete the plans as he had done with the GWR the year previously. Members of the committee visited Bridgwater, Taunton, Wellington and Exeter to gauge support for their scheme both with the public and local landowners. Buoyed by the positive response they received, they determined to press on. Brunel told the committee that completing the survey in such a short time would be 'impossible', particularly because of his other commitments, but that he would agree to be engineer because he had 'assistants in whom I can place perfect reliance and who by communicating to me the results of their surveys would render my directions in effect as well as I could myself'.[23] He added that he would send an Assistant Engineer who would be 'perfectly competent' and who with his assistance would complete the work as well as he would himself.

Although not named, this was William Gravatt, who had joined Brunel at the Thames Tunnel in 1826 when both men were just twenty. Gravatt and Brunel had worked closely together in trying circumstances in the tunnel and Brunel later recorded that they had been intimate friends, but after the flood that had nearly killed Isambard in January 1828, Gravatt's professional behaviour had been less agreeable, particularly after he had been passed over by Marc Isambard Brunel for the post of Resident Engineer in favour of the less experienced Beamish. In the aftermath of the accident and Brunel's recuperation and work in Bristol, Gravatt had worked in Yorkshire, but he returned south to work for Isambard on preparations and surveys for the GWR bill.

Within two weeks, Brunel was able to muster a group of assistants, clerks and no fewer than nineteen surveyors to undertake the work and he visited a number of places along the proposed line, in particular to decide on where the railway would cross the Blackdown Hills between Somerset and Devon.[24] Having ridden over the landscape, he identified two routes, the first via Tiverton, the second via Cullompton. The latter was agreed with Directors, and the survey was completed speedily with Gravatt, based in Wellington, managing the work. The plans, sections and book of reference for the seventy-five-mile line were ready to be deposited on 30 November 1835, a considerable feat considering the timescale and the winter weather the tired and bedraggled surveyors must have encountered tramping across the countryside that November.

While a number of landowners did raise objections to the railway, in general, opposition to the Bristol & Exeter scheme was less vociferous than that faced by the GWR during its parliamentary struggles. Brunel spent two days giving evidence to the House of Commons committee in March 1836, and with a few minor changes, the Bristol & Exeter Bill was passed without incident on 19 May 1836.The railway was to cost £1,500,000 and at the first general meeting of the company held on 2 July 1836 the Chairman Frederick Ricketts told proprietors that Directors were convinced of the 'adequacy of the proposed capital to meet any contingency' and that they believed the advancing prosperity of the country would make their investment worthwhile.[25] Brunel was thanked for his 'distinguished and unwearied exertions', and his appointment as Principal Engineer was confirmed. Referring to Gravatt, the Chairman noted that those who knew 'the extent to which the able and rapid survey in November last may be ascribed to his indefatigable energy and perseverance would be pleased to hear that he had accepted the post of Resident Engineer'. Although the construction of the line would be 'less expeditious' than the previous eight months work had been, Brunel hoped that the line from Bristol to Taunton would be complete in just under two years and the whole line in three-and-a-half years.

All therefore seemed straightforward for Brunel who, with luck, would add another railway to his growing collection of broad gauge lines without too much difficulty. In his 1839 guide to the line, Wyld had recorded that the Bristol & Exeter was to be built 'on the same grand scale as the Great Western, and under the direction of the engineer...who has distinguished himself so much by his exertions'. The country through which it passed was difficult, he concluded, but presented 'comparatively little cutting, embankment or expensive work'.[26] Once again, Brunel's unquenchable confidence was misplaced, and a combination of circumstances, some of which were out of his control, would mean that the Bristol & Exeter was not finally completed until 1844, more than three years later than planned.

Work began with a familiar entry in Brunel's diary for 28 June 1836; it recorded the 'First flag of B & E hoisted' as sections of line were laid out on the ground by surveyors. The route would according to the Act of Parliament run from a 'junction with the Great Western Railway in a certain field called Temple Mead... and terminate at or near the City of Exeter in certain meadows on the south-western side of the new Basin and Wharfs in the parish of St Thomas the Apostle'.[27] A new bridge was required to cross the River Avon, giving access to the station at Bristol, and from there the line was to progress through deep cuttings at Pylle Hill and Ashton. The Ashton cutting was originally planned with a steep 1 in 45 gradient, requiring assistance for trains from a stationary engine on one side and a banking locomotive on the other, but this was eventually re-engineered by Brunel to reduce the incline significantly. The line climbed gradually out of Bristol before descending for almost five miles to a stretch of railway almost thirty-two miles long that was largely flat, running through the Somerset countryside passing Yatton, Highbridge, Bridgewater and Taunton. From there the railway climbed the 'Somersetshire Alps', the Blackdown Hills, where the most significant piece of engineering on the line, the Whiteball Tunnel, was reached. At over 1,000 yards long, the brick-lined tunnel was driven through clay and sand. The railway then ran for a further fifteen or so miles through Tiverton Junction and Cullompton to Exeter. Branch lines were initially planned for Bridgwater and Tiverton, with extensions to Clevedon and Weston-Super-Mare being added in later years.

In addition to Gravatt, Brunel also appointed three assistants, William Froude, John England and R. A. Welsh, who were initially based at offices in Coronation Road Bedminster, then on the outskirts of Bristol. Although contracts were let to William Ranger and Samuel Hemmings at the end of 1836 for the first stretch of line from Bristol westwards, work proceeded slowly owing to both a financial recession and a cholera outbreak.[28] Initial enthusiasm for the project had not manifested itself in actual investment and the company found itself short of capital. In August 1837, Brunel was asked to make recommendations on how work could be phased, and at a board meeting soon after, Directors told shareholders that they would use the funds they had to the best effect by 'not operating on too large a surface at a time' so that 'investment and return may, in a good degree, keep pace with each other'.[29] Writing much later, Thomas Latimer reported that 'serious discussions had taken place amongst Directors as to whether it would not be the wisest course to wind up or abandon the undertaking.'[30]

Work continued slowly, and while large sections of the line had been laid out much of the land needed had not yet been acquired, presumably because of lack of money. A further Act of Parliament was therefore required in June 1838 to extend the time needed to complete the purchase of land, and despite the seemingly gloomy outlook the Company also asked permission to build further branches to Nailsea, Weston-Super-Mare, Bleadon and Burnham. To add to Brunel's woes, the difficulties faced by Ranger in fulfilling contracts on the GWR were also becoming apparent on the Exeter line. After some discussion, the engineer was able to negotiate a transfer of contract, which meant that after a further tender process Hennet took over the work, although Brunel was forced to accept a higher price for the contract. The severe weather over the winter of 1838 and 1839 that had caused so many problems on the Great Western also took their toll further west; with the financial position of the company still difficult during 1839, the B & ER Directors were forced to take more drastic steps and approached the GWR with the proposal that the larger company should lease their railway, opening up the possibility of access to more capital.

Some inkling of what might lay ahead came in the appointment of a new Managing Director, Charles Fripp, brought into the fold to address slow progress on the line. Fripp was already a Director of the Great Western Railway and brought a more robust approach and was clearly a strong link to the larger and more powerful company. His appointment was also perhaps an acknowledgement that Brunel's involvement in the GWR, the *Great Western* steamship and other projects did not allow him to concentrate as fully on the Bristol & Exeter as their Directors might have wished. There had already been debate in August about the possibility of opening as far as Bridgwater with a single line, but following further discussions the B & ER were told that this would cost £700,000, the addition of a second track adding a further £112,000. Brunel told the board in January 1840 that while only laying one set of rails would save money there were disadvantages; not only were there practical limits on the number of trains that could be run, but there was also the danger that opponents of the railway would seize on the move as a sign of weakness and promote the perception that the railway was not safe.

There does not seem to have been any direct mention of any potential conflict of interest Brunel may have had since he was engineer to both the GWR and the Bristol & Exeter. Indeed, in his comments to the board he argued that the disadvantages mentioned would be much reduced if the line was operated as part of the GWR. In the spring of 1840 with the railway still well behind schedule and with many investors unable to actually pay for shares, Bristol & Exeter proprietors were asked to allow the board to negotiate a lease with their bigger neighbours. After some debate the meeting voted to allow this to take place and an agreement was submitted for discussion at a General Meeting on 2 September.

Brunel as usual presented an upbeat assessment of the prospects for the railway if a lease was agreed, telling the meeting that the B & E would lay a double line between the two cities, and a single line branch to Weston. In return the GWR would lease the Bristol & Exeter for five years, paying the company £30,000 per year plus a farthing per mile for each passenger or ton of goods carried on the line to Bridgwater, with increases as the line was gradually completed.[31] As might be expected, there was a heated discussion about whether the smaller concern should go it alone, or allow the larger GWR to step in. There was a feeling that the Bristol & Exeter's original traffic figures had been overestimated by the promoters, something that in truth was not confined just to that company during this period, and with no consensus the decision was deferred to a meeting in September, when it was finally confirmed by shareholders with 3,550 in favour and 1,014 against.

With a more stable financial situation progress began to be made, and there was more positive news for Directors and shareholders at the end of May 1841, when the first stretch of line from Bristol to Bridgwater was finally opened. The thousands of people who flocked to the town on Monday 31 May were to be sadly disappointed as *Javelin*, the engine scheduled to haul the special train to Bridgwater, had been derailed and damaged the night before and no one had thought to let civic authorities know that the train and celebratory breakfast had been postponed to the following day. Festivities were resumed on Tuesday, and 400 guests travelled the thirty-three miles from Bristol in one hour forty-five minutes. Champagne and claret were served at a celebratory lunch before the train returned to Bristol.[32] Some further work required by Railway Inspectors meant that the line did not actually open to the public until 14 June.

While this good news was clearly welcome, Brunel had a further issue to deal with that must have added to the worries he already had, both on the Bristol & Exeter and the GWR. By the spring of 1839 it had become apparent that his relationship with William Gravatt was not all it could be. Although it seemed from the beginning that Brunel as engineer would retain responsibility for all aspects of the construction of the railway, his rapidly growing responsibilities necessarily entailed the delegation of some work to Gravatt, who had been so valuable to him during the initial survey. With Brunel preoccupied, Gravatt was soon seen by many as being the engineer in charge of works on the new line rather than 'assistant' or 'resident' engineer, a perception he did little to dispel. A local newspaper praised 'the well known talents of Mr Gravatt the engineer' and after a General Meeting on 14 August 1838 in Brunel's absence, forty shareholders were taken on a three-mile journey along the partially completed line by Gravatt. At a dinner held afterwards at Congresbury, to cheers from the assembled guests Gravatt's health was toasted.

In April the following year, the contractor Hemming wrote to the Bristol & Exeter board asking to be released from his contract alleging that he had been badly treated by Gravatt and his staff. Given the lack of progress made on the works, this was a serious threat, but Brunel was able to dissuade him from walking away from the contract. Writing to Gravatt in April 1839 Brunel asked him to keep a strong check on his staff and their 'unavoidable tendency to abuse their power' and to treat Hemming and other contractors fairly. Having read in earlier chapters of Brunel's poor treatment of contractors such as Burge and Peto, readers may perceive more than a little hypocrisy on his part, but it may be that he was also aware that the problems may have arisen because he should have paid more attention to the work of Gravatt and his staff in the first place. More significant, however, were his comments that 'Some change must also take place or rather return to the more careful observance of the relative position in which we stand'.[33]

While recognising that Gravatt did not always refer matters to him 'from a desire to save me trouble' Brunel instructed him to communicate more regularly, and more importantly that in 'communications with Directors you should avoid as much as possible advancing your own opinions upon new subjects until you have communicated with me'. This warning suggests that Brunel suspected that Gravatt was projecting an image of being more than his assistant or at least was not doing anything to prevent this view being more widely promoted, as an article in the *Somerset Gazette*[34] in November 1839 seems to confirm; the writer described him as 'the talented engineer of the company'. The following month Brunel wrote to Gravatt yet again after it appeared that he had been talking to Directors without permission, threatening him with dismissal. After some discussion the matter was smoothed over but in May 1840 matters came to a head when more evidence came to light that Gravatt had been in contact with board members and had been providing them with his thoughts on 'important engineering questions', which seemed to question the 'correctness' of Brunel's views. An emergency board meeting was held to discover more, and following this the engineer wrote to Gravatt asking whether his actions were 'the conduct of a friend, of a gentleman, of a subaltern trusted & confided in by the man above him'.[35] At a special board meeting held on 25 July 1840, Brunel's work was unanimously supported by the board, and it seemed that the only course of action was to sack Gravatt, but perhaps still clinging to some feeling of friendship, he instead instructed his resident engineer that he was to work only on the Bristol to Bridgwater section, adding that he was afraid that 'vanity' would prevent Gravatt from 'occupying the place of second to any man'.[36]

Despite another reprieve, Gravatt continued to overstep his authority. Concerns about the supervision work on various bridges which had settled or collapsed and the appointment of a pupil without his permission led Brunel to write again, revisiting his concerns about Gravatt's overstatement of his role and reminding him that in Bristol & Exeter matters 'you act as my assistant – and that everything done for which I am made responsible must be done in my name.'[37] When Gravatt failed to report that a bridge at Exeter was in a state of collapse and another at Bristol was in a 'deplorable state' Brunel's patience finally ran out, and he wrote to Gravatt telling him clearly with some regret that he felt that he had treated him with respect for nearly fifteen years, but that he had now lost confidence in him. Brunel offered Gravatt the chance to resign so that the matter might be treated as private, but this offer was not taken up, and on 18 June 1841 he sacked him.[38] The matter was not concluded there, as Gravatt then attended a board meeting arguing that his dismissal had been unfair and the result of a conspiracy. Angry letters followed but the board continued to back Brunel, who was clearly exercised by the 'ravings of a madman without a single fact or alleged fact'.[39] There was more trouble at the September General Meeting when Gravatt made an address to proprietors and a two-hour debate then followed. Brunel told the meeting that he should not be present while his own conduct was discussed, and as a newspaper reported, 'Mr Brunel put on his hat and left the meeting.' Gravatt with perhaps some truth argued that things might have been different if the GWR had been completed and Brunel had been able to spend time to 'attend the Bristol & Exeter'. A motion of no confidence tabled, accusing Brunel of negligence and misleading the Directors, which was rejected, and an amendment supporting him was carried to the sound of loud cheers.

Despite all this upset, work continued apace to complete the line. John England was appointed as Resident Engineer, and Brunel, with work on the Great Western Railway now completed, spent more time in the West Country, more convenient for the Bristol & Exeter Railway and the Bristol & Gloucester and Cheltenham lines, as well as being close to the work now been undertaken in Bristol to build his next steamship, the SS *Great Britain*. Brunel's office diaries for 1842 and 1843 continue to record the breakneck pace of his working life. In 1842 he had spent a good deal of the year based in Bristol and Weston-Super-Mare, along with other locations on the route of the Bristol & Exeter including Bridgwater, Taunton and Wellington, with many entries noting that he was working 'along the line';[40] although by the beginning of 1843 he was also closely involved in supervising the final completion of the SS *Great Britain* at the Great Western Dockyard in Bristol. On 3 April 1843 Brunel suffered another potentially life-threatening accident when performing a conjuring trick using a half-sovereign with some children at his London home. Placing the coin on his tongue he accidentally swallowed it. For several days he did not feel any ill-effects but travelling to Bristol on 7 April he felt unwell, and it soon became apparent that the half-sovereign might be lodged in his trachea or lung. After being examined by Sir Benjamin Brodie, Surgeon to the Queen, various attempts were made to dislodge the coin to no avail, including a tracheotomy performed on 27 April. The crisis was finally averted on 13 May when Brunel was strapped to a hinged table of his own design that put him in a prone position. With some persuasion the coin dropped back into his mouth; 'At four ½ I was safely and comfortably delivered of my little coin'[41] he wrote to Captain Claxton the same evening, adding that he hoped to be in Bristol by the end of that week. Although he made light of his escape, he was nevertheless persuaded to take some country air to recuperate and allow his neck to heal.

Brunel's accident had been a matter of some importance nationally with daily reports in newspapers throughout the forty-one days he was incapacitated; even the *Railway Times*, one of his fiercest critics, was prompted to offer good wishes, and reported the good news when the coin was removed. *The Railway Times* subsequently noted that 'Mr Brunel has now almost recovered his wonted strength' and it seems that he was soon back at work; in the summer of 1843 was based in the York Hotel at Clevedon. With some short interruptions, he spent much time there between June and the end of August catching trains from Yatton (then called Clevedon Road station) to Bristol and beyond, or west towards Bridgwater and Taunton.[42] On 18 July 1843 Brunel travelled back to London from Clevedon on the eve of the launch of the SS *Great Britain*. On the following morning Prince Albert joined Brunel and Gooch at Slough and then sped down the GWR main line to Bristol where he was greeted by thousands of people on the streets of the city and around the dockyard. Brunel returned with the Prince to London after the launch and in characteristic style then had further meetings at Duke Street.

The Bristol & Exeter line had reached Taunton on 1 July 1842, and work had proceeded well enough for Brunel to report to the board in September the acceptance of a tender from Bridgwater contractor George Hennet for the installation of the permanent way from Taunton to Exeter at a cost of £2,780 excluding rails and timber.[43] After eight long years, the line was finally completed to Exeter on 1 May 1844. Reporting to the board a few months later Brunel was able to tell them that the last forty miles of line, 'on the construction of which, many persons had anticipated great difficulties and expense', had been 'economically executed' and that the 'very great outlay' which the protracted construction period had entailed was still within the sum allowed for in the original Act. The Bristol & Exeter he added was one of 'the very few Companies in England which have not been compelled during the progress of their works, to return to Parliament for powers to increase their capital.' Without irony he concluded that he was 'far from taking credit to myself, for all these favourable circumstances but I refer to them as rendering my work on the present occasion a very agreeable and a very easy one'. Perhaps hinting at their support during the Gravatt affair, he thanked Directors for the assistance he had received, and the encouragement always given to him by the confidence they had placed in him 'under very peculiar and difficult circumstances and which confidence I trust you have not had reason to regret'.[44]

Brunel's triumphant report to the board in 1844 did not mention the fact that trains reaching Bristol were still using his Great Western Railway terminus, which involved an inconvenient and time-consuming reversal of trains in and out of that station. Arguments between the two companies about rent and lack of money had meant that Brunel did not present proposals for a station until early in 1845; an early sketch suggests that he wished to build a station very like his GWR train shed, but these ideas were rejected by the board and he was asked to design something simpler and cheaper. The result was a single storey timber shed with five tracks, built on vaults like its GWR counterpart. The two stations were linked by a double track curve later provided with an 'express platform' for through trains. Latimer was scathing about the arrangement: 'As if to give new illustration of the unpractical mind of the engineer, the station erected at Bristol was placed at right angles with the Great Western Terminus occasioning extreme annoyance to through passengers, and great delay.'[45]

The Bristol & Exeter station was completed within six months of tenders being placed in February 1845, but congestion in

Brunel's original station was only slightly reduced as trains from Gloucester and the Midlands had also begun using the terminus with the completion of the Bristol & Gloucester Railway the previous July. The railway, which Brunel had almost forgotten to mention in his 1835 Boxing Day diary entry, was a revival of an earlier proposal to link Bristol with Birmingham which had failed a decade earlier and also involved the absorption of part of the Bristol & Gloucestershire Tramway mentioned in Chapter Two. Brunel had made an early survey of a possible route in 1833 but did further work three years later when the Bristol & Gloucester Act was finally passed by Parliament. The proposed standard gauge railway would absorb the Bristol & Gloucestershire and extend its line from Coalpit Heath to Standish Junction near Gloucester where agreement with the Cheltenham & Great Western Union Railway then being constructed would allow a third rail to be laid to Gloucester itself. The twenty-one-and-a-half-mile line was to be built at a relatively modest cost of £400,000 with the Great Western Railway pledging £50,000 towards its construction. Brunel was offered the position of engineer on 9 September 1839 and the Chairman George Jones noted that the 'narrow gauge' would be adopted and also that the permanent way should not be the 'baulk road' being used by the GWR on its new line. This would mean that the railway company would be constructed 'at an outlay not exceeding the means at their disposal' he added.[46]

Despite all he had already said about the adoption of the broad gauge and the difficulties he had encountered with the Liverpool Party on the GWR in 1838, when accepting the Bristol & Gloucester job a few days later, Brunel raised no objections to these conditions, and may have been biding his time before raising the issue again. Proposals to amalgamate the fledgling railway with the Birmingham & Gloucester company to create a through route to the Midlands in early 1840, which also involved the potential purchase of the Cheltenham & Great Western Union Railway broke down and so negotiations with the Cheltenham company were eventually concluded with the promise that they would lay a third rail between Standish and Gloucester, although there was no firm agreement about when this work would be done, only that it would be the last section of their railway to be completed.

Brunel got his opportunity to argue for his 'pet crotchet' the broad gauge' soon after when Directors asked him to produce a special report on the implications of adopting the 7 ft gauge. His conclusions were that it would cost a further £6,968 to add an extra foot to the permanent way between Bristol and Standish. A final decision on the track gauge was not made until March 1843 when shareholders were told that the Cheltenham & Great Western Union Railway was to be taken over by the GWR. The Bristol & Exeter had also been leased to the Great Western in the years following the establishment of the Bristol & Gloucester company, and it now risked being isolated from potentially important sources of traffic, and the inconvenience for through passengers of changing trains at Bristol was also a consideration.[47] An agreement was signed with the Great Western that gave the Bristol & Gloucester running powers over their lines, at a cost of £11,000 per year, along with access to their passenger stations and goods depots at Bristol, Gloucester and Cheltenham, although also at a further cost of £3,500 per annum. The GWR agreed to complete the track between Standish Junction and Cheltenham by April 1844, saving the company around £30,000. In return, the Bristol & Gloucester agreed to complete its line as a broad gauge railway, making a connection with the GWR at Bristol and contributing £50,000 to the new South Devon Railway being actively promoted by Paddington to push the broad gauge even further west. The *Railway Times* suggested that the B & GR had bound themselves to the Great Western 'hand and foot, henceforth and forever' although as events would turn out, this would not be the case.

After a rapid survey by George Hennet the line had been quickly laid out and construction work had begun in 1841, so when a decision to adopt the 7 ft gauge was made two years later much work had already been done, meaning that the railway was not engineered with the clearances seen on other Brunel railways. The line was relatively short with few engineering difficulties; there were however, seventy-three bridges on the twenty-one-mile route, fifteen of which were constructed of timber. Helpfully, the geology was more favourable with shales rather than clays being common, making slips on embankments less likely. Directors and invited guests took a slightly hair-raising special trip from Bristol to Stonehouse on 13 June 1844, travelling in two coaches and an open wagon at speeds of up to 60 mph, and the railway was opened from Bristol to Gloucester on 6 July. It was an eventful day, with the inaugural train leaving Bristol over two hours late and then derailing itself just outside Gloucester. It was reported that the 500 passengers got off the train and walked the short distance to the Birmingham & Gloucester station where they enjoyed a banquet. Brunel, Hennet and the Inspector of Railways Major-General Pasley following in an inspection train behind missed the celebrations as they helped re-rail the stranded locomotive. Despite not being able to complete his journey and discovering that several miles of the line were only single track and that some stations were not complete, Pasley accepted Brunel's assurance that all would be rectified, and passed the line, returning two days after the start of public trains on 10 July 1844 for a further inspection.

At Gloucester, trains used two temporary platforms at the Birmingham and Gloucester station and goods and passenger traffic was healthy, although arrangements at Gloucester where broad and standard gauge services met did not prove to be the straightforward arrangement Brunel had promised. Although the opening of the Bristol & Gloucester established a through route between the south west and the north-east of England, problems caused by what became known as the 'Break of Gauge' created what one nineteenth-century commentator called a 'wall across the way' and brought about another watershed moment in Brunel's career.

Brunel had shown little enthusiasm for the proposed line from Swindon to Cheltenham when he noted that he had only agreed to be engineer 'because they can't do without me' and like the Bristol & Exeter its construction turned out to be more complicated than he had initially envisaged. It also demonstrated once again his propensity to take on more projects than he could realistically cope with. The passage through Parliament of the Cheltenham & Great Western Union Railway Bill in 1836 was by no means as easy as that of the Bristol & Exeter, for the most part because of the opposition of the London & Birmingham Railway, who feared the spread of Brunel's GWR into the Midlands and beyond. They had also supported the promotion of a rival line that would run from Tring to Cheltenham via Aylesbury, which would in addition include a branch to Oxford and into the Cotswolds. This proposal was eventually defeated, but rivalry between the two companies would continue throughout the rest of Brunel's career as a railway engineer and endure for some years after his death. The promoters of the Cheltenham line faced familiar opposition from canal owners, this time from the Thames & Severn Canal, and landowners including Squire Gordon of Kemble, someone we shall meet shortly.

The Act finally passed by Parliament also included details of arrangements with the Birmingham & Gloucester Railway who had recently received permission to build their line, and it was agreed that both companies would share a line between Cheltenham and Gloucester. Although the track gauge of the Cheltenham line was not mentioned specifically, since it would strike north from Brunel's GWR at Swindon, it would be broad gauge. The Act stated, however, that the line between Cheltenham and Gloucester would be 'formed in such manner and with rails of such shape and width as shall be convenient and adapted for the use of carriages running on the Birmingham & Gloucester and London & Birmingham Railways'. It was then noted that if any further rails were to be laid down on what would be a standard gauge stretch of railway, then the cost would be borne by C & GWUR.[48] The first prospectus reported that following a survey by Brunel, the railway was estimated to cost £750,000. An entry in Brunel's letterbook recorded that this sum included 'Excavations, Embankments, Tunnelling, Masonry, forming the Road, Purchase of Property and Contingent expenses'.[49]

The woes felt by the Bristol & Exeter company due to the economic slump were shared by the Cheltenham Directors, and there was little progress in the year following the passing of the Act. Having worked on the Great Western Railway, Charles Richardson had been sent to Gloucester by Brunel in September 1836 and helped lay out the route of the line from Cheltenham to Stroud, having already made a number of preliminary excursions to Gloucestershire with him the year before. After a brief stint at the Thames Tunnel, Richardson returned to work on the C & GWUR in January 1837, and following his trip to check on the production of rail for Brunel in South Wales he remained on the Cheltenham line until its completion in 1845.[50] Richardson's diary reveals the financial difficulties faced by the railway in its early days; as well as supervising the survey and excavation of test pits, Richardson also bought materials for the company who relied on credit from local banks and the goodwill of traders, but did not help themselves by the slow settling of accounts. In August 1837, Richardson noted that having completed his accounts, he had 'brought the C & GWUR Co & myself to have, each, about £18 in the bank'.[51] Within weeks he was told to cease work on the Gloucestershire section of the railway.

Unable to raise enough funds despite the initial enthusiasm shown for the project, in November 1837 the Directors made the decision to abandon work on the line over the Cotswolds to Cheltenham and Gloucester and instead concentrate on the line as far as Cirencester. This news was not well received by the Birmingham & Gloucester Railway, which was making good progress and had expected the Gloucester-Cheltenham line to be built first. When the Cheltenham company returned to Parliament in 1838 to ask for further time to complete the railway and vary the route originally planned by Brunel, opposition from the Birmingham & Gloucester resulted in an additional clause ensuring that the Cheltenham to Gloucester route should not be abandoned by the C & GWUR, and if necessary, the Birmingham & Gloucester were given powers to build it themselves, which in the event they were forced to do in 1839 when it became clear that the Cheltenham company were in no position to comply, the line being opened in November 1840.

Some idea of just how short of money the C & GWUR was is illustrated by the fact that by the end of 1839 only £200,000 of the £700,000 needed to build the railway had been raised. Little had been done between Gloucester and Kemble apart from the sinking of trial shafts for the tunnel needed at Sapperton and the focus instead shifted to the Swindon end of the route and the line from there to Cirencester. Completion of this stretch was originally planned for January 1841, but a landslip at Purton caused by wet winter weather slowed progress. Brunel instructed the contractors to burn clay on the slopes of the embankment that had slipped and then level with dried rubble and sand.[52] With this problem overcome, the first section of railway to Kemble with intermediate stations at Purton and Minety was completed on 31 May 1841. Since work had been paused on the main line north of Kemble, the completion of the short branch to Cirencester meant that for some years the small market town acted as the temporary terminus of the whole railway. Kemble, intended to be the junction for the branch, was not a public station and was only provided with two wooden platforms where passengers could change trains. This state of affairs was a consequence of a fractious relationship between the railway and the local landowner Squire Robert Gordon, that once again illustrated the premiums paid by railways to secure land they needed. Gordon, an avowed opponent of railways, had been a thorn in the side of the company during the parliamentary process but had nevertheless been persuaded to sell land at an extortionate cost of 7,500 sovereigns. He had also negotiated a difficult legal agreement that prevented the C & GWUR from not only building a station at Kemble, but also compelled them to hide the railway within a 415-yard tunnel so that steam and smoke from locomotives could not be seen from his residence, Kemble House.

On 27 June 1842, just over a year after the railway opened, Gordon complained that engine noise at the junction was causing a 'nuisance'; company solicitors tested his complaint by asking driver Jim Hurst, a regular on the Cirencester branch since its opening, to blow the whistle of the locomotive *Lion* while they listened from inside and outside the landowner's house. Although the wind was blowing from the junction, there was nothing to be heard and no smell of smoke either. There was a further legal challenge later in the year when objections were raised about the operation of the engine shed at Kemble, especially the use of a blacksmith's forge there. The company was able to continue using the shed for engine repair and cleaning but prevented from extending their operations.

Obstruction from the Gordon family meant that Kemble did not appear on GWR timetables until 1872, and a proper station was not completed until a decade later. In the interim a main line station was provided at Tetbury Road on the far side of the Fosse Way turnpike, which also had a large goods shed – that presumably handled agricultural produce from the Gordon estate.

E. T. MacDermot records that after the completion of the Cirencester section Directors were 'anxious to get rid of the whole undertaking and their responsibilities as soon as possible'[53] and in 1842 obtained a further Act of Parliament giving it power to 'complete the line through the aid and co-operation of other companies' and to raise more money as the original estimate was 'inadequate'. Although a section of the line had been completed and opened to the public, the construction of the remainder had been 'delayed from various causes not within the control of the C & GWUR', and it could not complete the railway in the time allotted in the original Act. Powers to raise a further £750,000 were sanctioned, although it was soon apparent that the Cheltenham company did not intend finishing the project. Negotiations rapidly took place with the Great Western Railway and in August 1843 C & GWUR shareholders agreed that the company should be sold to them for £230,000, a bargain for Paddington, since more than £500,000 had already been spent by the Cheltenham company.

The sale agreed, there was no time to waste and work started on completing the line from Kemble. Contracts were let for the construction split into four sections: Gloucester to Stonehouse, Stonehouse to Sapperton Tunnel, Sapperton Tunnel itself and Sapperton to Kemble.[54] Sapperton Tunnel, just over three miles from Kemble, proved to be a challenge for Brunel and the contractors. Driven through a ridge of the Cotswold hills dividing the Thames Head and the Golden Valley, Brunel had originally intended to excavate a single tunnel of around 2,800 yards, which would have been built on a long curve. This was judged to be unsuitable when the C & GWUR bill was debated in Parliament, and as a result the second 1838 Act proposed a straighter route, which Directors hoped would be cheaper and shorter. Preliminary shafts had already been sunk to check the geology, with Richardson spending much time on this task, recording in his diary that he had worked on drawings for the 'trial pits at Sapperton' at Duke Street in May and June 1837, before travelling to Gloucestershire and spending almost all of July at Sapperton supervising the work.[55] Despite further headings being started in 1841, nothing more was done until the railway was sold to the GWR, after which work began in earnest with the appointment of local contractor Nowell to excavate the tunnel.

Brunel had clearly had further thoughts about the line of the tunnel and more survey work was done, resulting in a third iteration of his design that now featured two tunnels. From the Stroud end, trains climbed the 1,855-yard 'long' tunnel at a gradient of 1 in 90, emerging into a short cutting, before plunging back into the 353-yard 'short' tunnel. It is thought that economy was the driving force behind this dramatic change rather than experimentation, but the result, while cheaper, was a tunnel that would be difficult to work, the steep inclines requiring banking engines for heavier trains until the diesel era. The Fullers Earth and Oolitic limestone found in the survey was familiar to Brunel and the navvies who dug it out, being similar to that of Box Tunnel. The completion of the contract seems to have been more straightforward, although there were still a number of fatal accidents involving navvies during construction, particularly during blasting.[56] It was reported that men worked day and night at times, but the tunnel was complete by February 1845, well within the length of the contract.[57]

From Sapperton, Brunel's railway snaked through the Golden Valley for almost seven miles before reaching Stroud, and then continued on through Stonehouse to Standish Junction. The Stroud Valley stretch was particularly characterised by the use of timber viaducts, utilised by Brunel on most his railways, particularly those in the West Country. There were nine timber bridges on the Cheltenham line[58] of various sizes and designs, a number crossing the Thames & Severn Canal that ran parallel with the railway for some distance at that point. Although these timber structures were an elegant and cheap solution, they left the railway with a financial headache in subsequent years and in the case of the C & GWUR this problem came to light as early as 1858 when Brunel was forced to admit to Directors that £35,000 was needed to renew timber viaducts, many of which were in the Stroud Valley.[59]

By May 1845 the railway was complete and it opened for passenger traffic on 12 May. *The Gloucester Journal* reported the event noting that the completion of the line finally enabled local people to travel to London within four and a half hours, with an 'express' service reducing that journey to just under three hours. The complicated arrangements at Gloucester necessitated two junctions enabling traffic from the Birmingham & Gloucester, Bristol & Gloucester and the GWR to run safely, while additional rails had also been laid by the GWR on the much debated Gloucester to Cheltenham section, enabling Brunel's broad gauge trains to access their station at Cheltenham St James.

7

Broken Dreams

On a chilly winter's evening in January 1845, the Albion Tavern in Bristol was packed with subscribers and well-wishers who had gathered to pay tribute to Isambard Kingdom Brunel, engineer of the Great Western, Bristol & Exeter, Bristol & Gloucester and Cheltenham & Great Western Union Railways. Brunel was the recipient of a grand silver-gilt dinner service dominated by an enormous ornate centrepiece in the form of a candelabrum decorated with figures representing 'Science, Genius and Invention aiding Commerce', three dessert stands, two sideboard dishes, salt cellars and silver spoons. The generous gift had cost more than two thousand guineas, and had been paid for by more than 250 subscribers and the Great Western Railway.[1]

Before sitting down to a dinner described by *The Railway Times* as having 'the comfort, quiet and elegance of a private party', there were 'deafening cheers' from almost a hundred subscribers as the testimonial gift was unveiled. The evening was chaired by Charles Russell MP, Chairman of the Great Western Railway, and after the usual loyal toasts, it was Russell who proposed a toast to the health of the distinguished guest Mr Brunel and paid tribute in a short speech to his ability and energy and the successful completion of the Great Western Railway and other projects. Russell's toast was returned, *The Railway Times* reported, 'three times three' and followed by more cheers. Brunel seemed taken aback by this outpouring of praise and affection from those gathered at the Albion Tavern that night, and in his reply the normally bullish engineer admitted that he found it difficult to adequately return thanks to all his friends and colleagues around him, and that the gratitude he felt 'amounted almost to a feeling of pain', but he would never forget that day. Sitting down to yet more cheers, the dinner did not finally begin until more toasts had been made to Brunel's father Marc, his wife Mary and the Chairman and Secretary of the GWR.

When the dinner and presentation took place Brunel's career was indeed at a high point; in the four years following the completion of the Great Western Railway, the Bristol & Exeter, the Bristol & Gloucester and other lines which would form part of his broad gauge railway empire had also been opened with others under construction. Although work on the Clifton Bridge was progressing slowly, Brunel had also found time to design and launch two ground-breaking steamships, the *Great Western* and *Great Britain*. After the cheers at the Albion Tavern event faded, Brunel returned to London along his new Great Western Railway; his rise seemed unstoppable. The year 1845 would however be a turning point, and while there were still triumphs to come, the years ahead would prove to be more challenging than he might have imagined after the glow of that memorable winter evening in Bristol.

On 17 June 1845 Brunel had yet another lucky escape. He was travelling on a train from Paddington which was derailed near Slough while running at a speed at more than 60mph. Although the engine remained on the track, all five carriages in the train came off the line and were violently thrown down an embankment. A newspaper report gave details of a 'miraculous escape' for the 130 passengers, although there were a number of serious injuries suffered by travellers. Bruised, shocked but uninjured, Brunel, one of his assistants and Seymour Clarke, the GWR Superintendent of the Line, were able to help casualties at the scene and escort them to the Royal Hotel at Slough where first aid could be administered.[2]

Back at Duke Street, Brunel's office had never been busier; in 1844 he had been appointed as engineer to the Berks & Hants, Monmouth & Hereford, Oxford & Rugby, Oxford, Worcester & Wolverhampton, South Wales and Wilts, Somerset & Weymouth companies, and the South Devon Railway, the next link in the westward chain of railways after the Bristol & Exeter had received its Act of Parliament, allowing work to begin there. He had also been busy with work on railway projects in Italy.[3]

These appointments coincided with the onset of what became known as the 'Railway Mania', a frantic period of activity when railway speculation reached fever pitch, driven by a general belief that huge financial returns might be generated from investment in any new railway scheme. This was encouraged by unscrupulous railway barons such as George Hudson, who used the savings of ordinary investors to develop schemes, many of which were unlikely to ever pay dividends.[4] Some schemes were never likely to be built, and only served to manipulate the markets or serve as funds for other speculation. Brunel was scathing about railway speculators, describing them as 'stark, staring, wildly mad' and wrote 'I do not intend to go mad and I soon should at the rate others are going.'[5] This wild speculation would, he felt, siphon valuable funds away from his own projects which he thought were sound investments with good foundations, although as Adrian Vaughan notes, he was not averse to undertaking surveys and estimates for schemes that might never be constructed to frustrate rivals when necessary.[6]

At the end of 1844 there were already 2,235 miles of railways in operation, with another 855 miles of line authorised by Parliament that were either under construction, or about to be constructed. H. G. Lewin noted that these totals represented the work of more than one hundred companies, although four railways dominated: the Great Western, the Midland, the London & North Western and the London & South Western. The Midland and L & NWR had only recently been created through various mergers with the London & Birmingham and Grand Junction railways but these and the London & South Western Railway emerged as the main protagonists in a battle with Brunel and the GWR as all fought to protect their own interests by either building railways, or supporting proposals for other lines from other associated companies that could act as feeders.[7] There is no doubt that the three companies ranged against the GWR would have already been formidable adversaries without the seemingly irreconcilable problem of Brunel's broad gauge providing an additional line of attack for them and transforming the issue from not only a battle for territory and traffic, but also a battle for the very future of his broad gauge dream.

Battle had been joined in January 1845 when the Bristol & Gloucester and Birmingham & Gloucester companies had amalgamated; discussions had then taken place between the newly created Bristol & Birmingham Railway and the Great Western Railway with a view to using the route to extend the broad gauge to Birmingham, but Charles Saunders had told his counterparts that the GWR would not negotiate further on a deal they felt already offered enough to the Bristol & Birmingham. John Ellis, deputy chairman of the Midland Railway, by chance then met two Directors of the Bristol & Birmingham returning to London after these unsuccessful discussions and suggested that if no deal with the GWR was possible, he would speak with them. Having already had some experience of Brunel's 'break of gauge' at Gloucester where passengers were forced to change from the broad gauge Bristol & Gloucester to the 'narrow' gauge Birmingham & Gloucester Railway, Ellis felt that the broad gauge had 'come too near'[8] and it was in his interest and of the Midland Railway to act, and act quickly. So it was that the Midland quietly acquired the Bristol & Birmingham line, by offering a better deal of six per cent on shares and taking over liabilities amounting to almost £500,000, a great bargain for shareholders of the Bristol line. Sekon argued, with some justification, that the GWR had become complacent and appeared to feel that the Bristol & Birmingham was virtually their own line anyway, and as a result did not pay sufficient attention to the importance of the negotiations. The deal was quickly ratified by the Midland board headed by Chairman George Hudson on 8 February 1845, and with it the first skirmish of the Battle of the Gauges was won by Brunel's opponents. The GWR could have little cause for complaint as the Midland had not acted in an underhand manner. Ellis summarised the fight ahead, arguing that the Great Western Railway 'had drawn the line themselves of demarcation between the two gauges; that of Oxford and Bristol are the places where they ought to change',[9] he concluded.

Clearly, Brunel and the Great Western thought differently and in the months and years ahead Gloucester became the place where arguments for and against Brunel's broad gauge would be vividly demonstrated, and its operation would be used by supporters and opponents of both gauges to argue their case. Most travellers on those lines felt having to change trains was a real inconvenience, however well station staff might deal with the process. One opponent writing in 1846 about their experience addressed the great engineer himself, asking whether 'Mrs Brunel herself' would think of making a journey at night and what she might think of changing trains 'with her two children, servants and the usual amount of family luggage'. Arriving in darkness, the writer noted, children would need to be woken and 'muffled up' against the cold before stepping from one carriage to another. Servants were 'not in the best of humour', and leaving the carriage, every corner had to be searched for 'bonnets, stray gloves, handkerchiefs, smelling bottles, sandwich cafes, baskets, parasols, umbrellas and unutterable small parcels'. Apart from providing a fascinating insight into Victorian train travel, this description also highlighted the delay, inherent chaos and hustle and bustle resulting in the break of gauge.[10]

As important were the implications of the break of gauge for the rapid transit of goods traffic. A revealing insight into the process was given in a letter to the Directors of the GWR printed and published in Manchester in 1846. The author, described only as 'An Old Carrier', was a trader who ran a goods business hiring a large number of railway wagons. He noted that railway companies for the most part worked well, but there was one part of the country 'that beats me completely and promises at times to drive me crazy'. To every other part of England he could generally guarantee delivery times, 'except beyond Gloucester'. He neatly describes the arrangements at the goods station; transhipment sheds, platforms, cranes, offices, clerks, men and horses. 'Two sets of wagons, two engines, two engine drivers, two brakesmen, and TWO SETS OF EVERYTHING have to be used instead of one,' he complained. It took an hour to transfer the contents of one wagon of miscellaneous merchandise to another, often with the result that packages were turned 'topsy turvy' or pitched 'higgledy piggledy' from the broad gauge to the narrow gauge wagon. The carrier went on to catalogue the casualties resulting from the transfer of every kind of goods from crushed fruit to broken furniture. The carrier ends his list of complaints by mocking the size of Brunel's goods wagons, arguing that they were so big that they were like 'employing buffaloes to carry a mouse' or 'conveying a chest of tea in Noah's Ark'.[11]

The battle was recommenced in 1844 with the promotion of two new broad gauge lines that were to run north of Oxford, a location only linked to the GWR itself the previous year. The Oxford,

Worcester & Wolverhampton and Oxford & Rugby railways were a direct threat to rivals as they probed deep into country previously seen by what supporters of Brunel's railway belittled as the 'narrow gauge'.[12] The prospectus of the Oxford, Wolverhampton & Worcester Railway (OWW&WR), issued in September 1844, made much of the economic benefits to be derived from the line, noting that it would provide 'direct railway conveyance' for the produce of the extensive collieries, iron manufactories and other industry in South Staffordshire and Worcestershire supporting a population of upwards of 200,000 people in that region. Adding that the railway would benefit agriculture as well, Directors 'confidently asserted' that there was 'not a district in the Kingdom now unoccupied by railway which affords so great a prospect of renumeration'.[13] The second new line was the Oxford & Rugby Railway, a 50-mile route via Banbury through which the Great Western would eventually gain broad gauge access to Birmingham via the Birmingham & Oxford Railway, which ran into Birmingham from Fenny Compton.[14] The Great Western Railway Chairman Charles Russell later intimated that the company had not originally intended to compete directly with the London & Birmingham Railway in building a link from London to Birmingham, but the purchase of the Bristol & Birmingham line by the Midland had changed matters.[15]

The 1844 OW&WR prospectus had noted that surveys were being undertaken 'under the superintendence of Mr Brunel', implying that he had spent less time traversing the route of the line as he had done on other projects and had left the physical work of the survey to his assistants. There had certainly been little time in 1843 to undertake such work as he had been engaged on the Bristol & Exeter, Cheltenham and Taff Vale railways as well as the SS *Great Britain*. His appointment diary for 1844 has only two specific mentions of the 'Worcester Line', a meeting at Wolverhampton on 17 May and a meeting with OW&WR Directors on 13 September.[16] Brunel provided an estimate that suggested the railway could be completed for £1,500,000, a sum that once again would prove to be hopelessly inadequate, and the railway would find itself short of funds from the outset. In April 1847 a report from the Commissioners of Railways noted that 'a considerable number of railways under the control of the Great Western Railway belong to companies nominally separate and independent, but are in fact connected with, or more or less under the influence of, the Great Western Company,'[17] and so it was with the Oxford, Worcester & Wolverhampton. The prospectus had told prospective investors that an arrangement had been made with the GWR by which they would provide £750,000 of capital in return for a rent of three and a half per cent plus a share in the profits. This agreement also entitled them to nominate Directors, and so the familiar names of Thomas Guppy, Frederick Barlow, Henry Simonds and William Tothill were all listed on the original management committee and went on to become Directors. The backing of the Great Western appeared to strengthen the case for the new railway greatly, but in the coming years what had at first seemed a convenient arrangement became a source of tension and disagreement amongst the OW&WR board, who came to resent the influence of their partners.

There was no doubt that there was a straightforward financial imperative for opponents of the Great Western Railway, particularly in the case of the London & Birmingham, which stood to lose most from the success of either the Worcester or Oxford & Rugby railway schemes. It responded by promoting a rival line, the London, Worcester & South Staffordshire Railway, and this and the two railways promoted by Brunel and the GWR were first considered by a panel of five commissioners appointed by government to try and make some sense of the flurry of speculative railway schemes then being promoted. Inevitably, the arguments soon turned to the suitability of the broad gauge and commissioners listened to much evidence about the inconveniences and delays caused where broad and standard gauge trains met. Brunel, appearing on behalf of the OW&WR, suggested that wagons with removeable bodies, a forerunner of modern shipping containers, could be used to speed up transhipment from broad to narrow gauge and vice versa. Unconvinced, the Commissioners report issued in January 1845 ruled in favour of the London & Birmingham scheme mainly on the grounds of gauge and repeated the view already promoted by rivals of Brunel that breaks of gauge should be at Bristol and Oxford, not Rugby or Wolverhampton.

Undeterred, Bills for both Brunel's broad gauge schemes were submitted to Parliament in the spring of 1845 with the Oxford & Rugby bill reaching the committee stage in March and the OW&WR

bill being read for first time in April. Both bills were then subjected to detailed scrutiny over a period of nineteen days, with more than 100 witnesses being called including Brunel, Gooch, Hudson and George and Robert Stephenson. Austin, representing the London & Birmingham, did not, Sekon later reported, 'forget to continually remind the committee that this question of gauge was the great one the committee had to settle'.[18] Amongst the railway engineers called in support of the London & Birmingham standard gauge line were the familiar figures of Wood and Hawkshaw who, the promoters of the OW&WR recorded, had been employed by supporters of the narrow gauge to report against the broad gauge and had 'failed to induce the general body of proprietors to abandon that principle'.[19] After hearing from opponents, evidence was then presented in favour of the GWR and the broad gauge. Brunel seems to have demonstrated the same level of resilience he had during the committee stages of the GWR bills, and spent considerable time once again being cross examined. *The Railway Times* reported that as well as answering questions from the promoters' and opponents' counsel, he also described in detail the engineering features of the proposed line. In his evidence, Brunel reminded the committee that the OW&WR had been promoted as long ago as 1836 and that it was shorter by 15 miles than the rival scheme and most importantly, it ran across a far more densely populated area than that of the proposed London & Birmingham line. In answer to questions about the break of gauge, he repeated that he saw no difficulties arising from the transhipment goods wagons from one gauge to another as 'he had invented a simple "contrivance" to carry out this purpose, and it worked admirably'.[20]

When the committee finished its deliberations after more than 12,000 questions to witnesses and weeks of evidence, the chairman stated that they had heard quite sufficient in favour of the broad gauge schemes and that this part of the case was closed. In spite of further ferocious opposition from the rival railways and politicians such as Richard Cobden, Bills for the Oxford Worcester & Wolverhampton and Oxford & Rugby railways were given their third reading on 24 June 1845 and sent to the House of Lords. Once again, Brunel faced detailed questions in more committee proceedings over a number of days and spent yet more time arguing for the railway and the broad gauge. One of his bitterest opponents, George Hudson told the Lords that however ingenious Brunel's apparatus for transhipping goods might be, he deemed it 'most inadequate to overcome the great and numerous evils of "break of gauge"'. He was strongly opposed to the introduction of the broad gauge into the district and considered the existence of two gauges in the country as a 'great evil'.[21] Both bills were finally given royal assent on 4 September 1845.

Brunel's apparent victory for the broad gauge was short-lived. Within days of the bills being referred to the House of Lords, Richard Cobden, a radical MP, advocate of free trade and avowed opponent, moved an amendment in the House of Commons asking for a Royal Commission to be convened to inquire into the whole question of gauge. Cobden argued that following representations made to him from traders and manufacturers, 'serious impediments to the internal traffic of the country are likely to arise from the breaks that will occur in railway communications from the want of a uniform gauge.' The motion was passed and an announcement was made shortly after of the appointment of Sir Frederick Smith, the first Inspector-General of Railways, George Airey the Astronomer Royal, and Peter Barlow, Professor of Mathematics at the Royal Military Academy at Greenwich, as Commissioners. Their remit would be to investigate whether in future Acts of Parliament for railway construction, provision should made for securing a 'uniform gauge' for the whole country and also to decide whether it was practical to bring the railways already constructed, or those under construction, into line with a standardised gauge. Sensibly, the Commissioners chosen by the government had no railway experience or particular view on the question and it was hoped they could come to a fair decision.[22]

The *Railway Times*, not normally a supporter of Brunel or the GWR, reported the impending approval of both bills in August 1845 and argued that a 'terrible lesson' had been read to the London & Birmingham 'would-be monopolists' adding that their 'boasted anticipations of boundless power are sadly 'cabin'd, cribbed, confined' and that they should 'take care of what is left to them, rather than devote new modes of plundering their neighbours'. Telling readers that the debate between the rival companies had not been about gauge but about territory, it concluded, de haut en bas, 'We trust to hear now more about the gauge.'[23] The pages of *The*

Railway Times would soon feature little else, and the gauge question would come to dominate the railway world for months to come.

The Commission began its work in August 1845 and continued until December, taking evidence from forty-eight witnesses including engineers such as Brunel, Gooch, Locke, Stephenson and Vignoles, and those in the business of developing and promoting railways such as Hudson, Huish, Saunders and James Ellis. Inspectors from the Railway Commission were also interviewed, as were Army officers questioned about the effect of breaks of gauge on troop movements. Evidence was heard first from opponents of the broad gauge. Joseph Locke asked whether he thought 'a diversity of gauge' was 'an evil' replied that he did, and that it was better to 'continue with the narrow gauge and have the double (broad) gauge upon the shortest length'. He also argued that the easiest solution would be to remove the broad gauge entirely, so that 'the evil which you are sitting here to consider would be in the best and cheapest way got rid of'.[24] Locke did agree that broad gauge trains were faster than those then seen on the narrow gauge, but attributed this to the superior gradients Brunel had been able to engineer on his lines; he also admitted that heavier trains could also be run on the broad gauge.

Robert Stephenson, described as a 'Civil Engineer, Manufacturer of locomotives, Coal Mine Owner and a Practical Miner', a personal friend of Brunel, was nevertheless unyielding in his criticism of the broad gauge, noting that not only did it not have any advantages over the narrow gauge, but it had 'several disadvantages' including the additional cost of construction, which, he argued, required embankments, tunnels and cutting to be at least four feet wider.[25] Engines, tenders and rolling stock were also more expensive he added, although there was no difference in coal consumption. George Hudson, who told the commissioners that he was the Director of various companies that amounted to between 800 and 1,000 miles of railway, stated that he was 'perfectly satisfied that everything is accomplished by the narrow that is accomplished by the broad', and since economy was so important, 'the Narrow Gauge, I should say was the better of the two.'[26]

Daniel Gooch noted in his memoir with some irritation that Brunel and Charles Saunders had gone abroad for a holiday in the autumn of 1845 leaving him to represent the interests of the company and the broad gauge. It was 'a responsibility I did not much like, but it was to be, so I undertook the task'.[27] The timing of this trip seems inopportune given the serious nature of the Commission, and when Brunel was finally called to give evidence on 25 October 1845, it was observed that he had little in the way of notes having just returned from the continent and seemed less well prepared than Gooch, who appeared later. He responded to more than 200 questions displaying a mix of bravado, defiance and contradiction. He set the tone when asked if having seen other railways in action since the opening of the GWR, whether he would still choose the broad gauge. To answer the question candidly Brunel replied, was to risk 'being accused of adopting wild notions...I should rather it be *above* than under 7 feet now if I had to reconstruct the lines.'[28] Asked whether he considered that 'serious impediments' in railway traffic would develop because breaks of gauge would cause a 'want of uniformity' he replied that 'some inconvenience will occur' but felt that if a network of railways over England was created, he thought it impossible that 'passenger carriages can be running in all directions over that network without changing' and that he did not think that it would be 'for the advantage of the public that they should'.

For someone with so much vision and a man who had indeed imagined an integrated transport system linking London and New York via his new railway and steamships, it seems strange that Brunel was unable to see his broad gauge network as no more than a regional system where passengers and goods could transfer at particular stations on to the rest of the national railway network. Challenged again later on this subject, he reiterated his belief that public advantage would be 'injured' by the adoption of a uniform gauge for the whole country.

He continued arguing that 'the spirit of emulation and competition kept up between great railway interests' both in terms of comfort and speed, had done more good than 'the uniformity of system which has been much talked of the last two or three years'. Adrian Vaughan has rightly noted that this argument seems weak considering that it was the quality of Brunel's well-engineered line and the improved quality of Gooch's engines that had brought about so much improvement on the GWR, rather than competition from other companies.[29] The Commissioners also questioned Brunel on his apparent inconsistency in adopting other track gauges on other railways he had engineered, in particular the Taff Vale Railway and the Turin to Genoa line, both built to 4 ft 8 ½ inch gauge, and railway projects in Ireland, utilising a 5 ft 3 in gauge. His response to these awkward questions was to argue that one of the main advantages of the broad gauge was that its trains could run at high speeds over long distances, and since in the case of the railways mentioned speed was not a priority, the broad gauge need not be adopted in those cases.

Before the end of proceedings, Brunel and Gooch were able to persuade the Commissioners to instigate a series of trials to demonstrate the relative worth of broad and narrow gauge locomotives. The unfortunate Stephenson standard gauge *Long Boiler* engine No.54 chosen for the task proved no match for the broad gauge Gooch locomotive *Ixion*; it could not match the *Ixion*'s top speed and managed to derail itself on one journey. Data from these trials and a deluge of other statistical information were added to the conflicting testimony of the witnesses and the Commissioners retired to consider their verdict. On 21 February 1846 *The Railway Times* reported that the 'long-awaited' report of the Gauge Commissioners had 'at length made its appearance', and that the substance of its 'twenty seven long-winded pages' amounted to a verdict completely in favour of the narrow over the broad gauge[30] but also peevishly argued that while there was much 'sound reasoning' in the report, there was also 'no less that amounts to mere twaddle'.

The report clearly made some effort to ameliorate the bad news it was delivering to Brunel and his broad gauge supporters. Before giving details of its findings, the commissioners noted that they could find little fault with the broad gauge itself, and that in terms of safety, speed and 'the convenience of passengers' it was in many ways superior to the narrow gauge, and that the public were 'indebted to the genius of Mr Brunel and the liberality of the Great Western Railway' for the higher speeds, and more generous size of railway carriages they enjoyed. *The Railway Times* was not impressed; while acknowledging this 'high compliment' their editorial reflected that it looked very much like a 'soothing plaster to the author and supporters of the very system which the Commissioners are labouring to abolish'. While the performance of broad gauge trains was undoubted and had been proved in the tests undertaken the year before, they were for a comparatively small number of passengers the report concluded, and the Commissioners believed that this was not as important as 'the general commercial traffic of the country'.

Despite the undoubted theoretical advantages of the broad gauge and the fine performance of Gooch's locomotives, the Commissioners report highlighted two issues on which the argument was lost. The first was the stark reality that amongst the huge amount of statistical information considered, by July 1845 only 274 miles of broad gauge railway had been built compared to 1,901 miles of narrow gauge. Given this situation, and the fact that it would not be economic or realistic to widen existing narrow gauge lines to accommodate the broad gauge,[31] the Commissioners found it hard to justify anything other than adopting the 4 ft 8 ½ in gauge as a national standard.

The second issue remained the controversial question of the break of gauge. Gooch had written that as far as the 'evil of brake (sic) of gauge' the GWR had a 'weak case'.[32] While much time and effort had been expended by Brunel and others in suggesting solutions, especially in the transfer of goods from one gauge to the other, these had failed to convince the Commissioners; Gooch himself was sceptical, noting that he 'never had any faith in any of these plans as workable in practice'.[33]

In his 1870 biography, Brunel's son Isambard wrote that the 'adverse report' was a great surprise to broad gauge supporters, implying that rumours had 'led them to hope for a different result'.[34] George Sekon, a firm supporter of Brunel, added that they 'expected a favourable report extolling the greater excellence of their system and for a clear and unreserved commendation of the 7 ft gauge...together with full leave for the extension of the Broad Gauge lines in all directions'.[35] This expectation was misplaced, perhaps a reflection of both complacency on the part of the Great Western Railway, and in the case of Brunel, a firm conviction that whatever his opponents might think or say, his broad gauge concept was correct, and the way forward.

Incensed by the Commissioners' verdict and what he saw as inconsistences and inaccuracies in the report, Brunel responded with a belligerent fifty-page defence of the broad gauge entitled

'Observations on the Report of the Gauge Commissioners'. Its conclusions repeated familiar Brunelian arguments; the questions about the break of gauge were a pretext for the idea of monopoly, and that even if there was a single national railway network, through trains would still be impracticable. Competition between the two systems was an advantage and not a disadvantage, he added. The Commission had ignored evidence proving that the broad gauge was safer and faster, and the experiments carried out had 'demonstrated beyond all controversy the complete success of the broad-gauge system'.[36] For these and many other reasons, it was he argued that there should not be any 'legislative interference' with the broad gauge system.

Brunel's 'Observations' were printed and circulated to every member of Parliament and far from ending the Battle of the Gauges, the publication of the Commissioners report prompted a burst of further pamphleteering with various tracts being published, many opposed to Brunel and the broad gauge.[37] Addressing Brunel as 'the inventor of the broad gauge system' and someone 'animated with both honourable ambition and an honest desire to serve the public well in your vocation' one writer, while praising his achievement nevertheless noted that 'travellers preferred certainty and comfort of transit to extreme speed' and that 'uniformity of gauge was the way to achieve 'greater perfection than hitherto'.[38] The 'Old Carrier' noted earlier who had described the difficulties he had experienced at Gloucester by the 'evil' break of gauge argued in a letter to GWR Directors published in Manchester in 1846 that 'delusive statements' about the break of gauge had been put forth by two or three gentlemen said to represent the Great Western Railway.[39]. 'L.s.d.' another anonymous critic, claimed that the 'broad gauge partisans' should, following the result of the Commission 'retire amid a flourish of trumpets' and that Brunel who had established a reputation that rested on more than 'so many inches of gauge' would suffer a loss of face.[40]

The tone adopted by 'L.s.d' was not always typical of some of the sensational and often hysterical reporting which characterised the Battle of the Gauges and often focussed on personal attacks and territorial claims rather than a discussion of the merits of the respective systems. The debate continued unabated until the report of the Commissioners was finally considered by Parliament in the summer of 1846. The lobbying done by Brunel does seem to have had some effect, and with the passing in August of 'An Act for the Regulating the Gauge of Railways' some of the conclusions of the Royal Commission were watered down in the final Bill, providing the GWR with a way forward from what might have seemed an impasse. Although the Act stated that 'it would not be lawful to construct a railway for the conveyance of passengers on any other gauge other than four feet 8 ½ inches' it also recorded the 'Exception of Certain Railways', which included a long list of Brunel's projects including the Oxford, Worcester & Wolverhampton, Oxford & Rugby, and South Wales railways and any railway 'Southward of the Great Western Railway' in the counties of Cornwall, Devon, Dorset or Somerset for which any Act was then going through Parliament, or was under construction. Brunel's broad gauge survived, but only just; his railway empire could continue to grow, but as some have observed, this outcome may well have been a hollow victory. Uniformity of gauge would finally come in 1892 and it would be left to his successors to eventually deal with the final demise of broad gauge[41]. By that time, Brunel had been dead for more than thirty years, and the Great Western Railway would count the cost financially for many years to come.[42]

Two years after the passing of the Gauge Act, Brunel was to suffer another setback and what many regard as one of his most conspicuous failures, the Atmospheric Railway. In 1836, at a point when he was beginning work on the GWR, Bristol & Exeter, Taff Vale, Bristol & Gloucester and Cheltenham railways, he also found time to survey the route of another link in the chain of his growing railway empire in South Devon. Early plans for what became known as the South Devon Railway running from the B & ER station in Exeter to Plymouth involved two possible routes, the first a line running through the South Hams, the second running along the coastline between Exeter and Newton Abbot but keeping inland behind Dawlish and Teignmouth to avoid the sea wall in those places. A number of large and costly bridges would have been required to cross the rivers Teign and Dart and with costs likely to be high, the project was dropped through lack of investment locally.

Another scheme surveyed by the engineer J. R. Rendel was proposed in 1840, which also failed to generate enough interest, but there continued to be calls for a new line particularly from the business community in Plymouth. Following Brunel's close call with the half-sovereign in May 1843, he had travelled to Teignmouth to rest and a local newspaper reported that he had arrived in the town and had taken a house near the beach where it was hoped that the 'salubrious breezes of this delightful watering place will contribute much towards his perfect recovery'.[43] It seems unlikely that Brunel would have spent all his time at Teignmouth resting and he may well have used the break to think further about plans for the railway. With the completion of the Bristol & Exeter imminent, there was revived interest in the South Devon Railway (SDR) scheme and by the autumn Brunel was back in Devon, his diary recording that he stayed in the *London Inn* at Dawlish on the 4 September, spending several days there and further along the coast with Charles Saunders. For the rest of the month, apart from two brief trips back to Duke Street, the engineer spent a good deal of time at locations along the route of the proposed railway such as Newton Abbot, Plymouth and Totnes. On 12 September the diary noted that Brunel had 'spent the whole morning discussing difficult points to Ivy Bridge' and ten days later he was at Plymouth with Lord Morley, where he went through his park to 'marshes to view where the railway line will come across the land'.

Brunel returned again some weeks later and on Saturday 21 October he met Directors of the 'Plymouth Railway' at the Royal Hotel in Torquay, having spent a busy week 'on the line' completing the new survey of the route. He then took the train to Teignmouth and Exeter, before returning to London on the Mail Train from Beam Bridge.[44] Support for the railway had been bolstered by backing from what became known as the 'Associated Companies', a group headed by the GWR, which also included the Bristol & Gloucester and Bristol & Exeter who were anxious to push Brunel's broad gauge railway west. An agreement was brokered that provided financial support, along with the right to appoint Directors to the new company, then called the Plymouth, Devonport and Exeter railway.

A prospectus was issued and an application was made to Parliament for an Act in the next session.[45] There were no opposing schemes promoted and relatively minor opposition to the bill in comparison to some of the railways Brunel had engineered previously.[46] In Teignmouth a number of 'dwelling houses and cottages' were listed in the schedule for land acquisition, but the SDR argued that the removal of these 'lower-class' dwellings would have a positive benefit, 'a healthiness to the inhabitants from free circulation of air'.[47] By the time the bill was passed by MPs on 4 July 1844, the rather clumsy company name had been changed to the South Devon Railway and its contents clarified the role of the Associated Companies. The formation of the railway 'would be beneficial to the interests of the Great Western Railway Company, of the Bristol & Exeter Company and the Bristol & Gloucester Railway respectively'. It noted that the GWR had subscribed £150,000, the B & ER £200,000, and the B & GR £50,000 and that there would now be twenty-one Directors, eleven of whom were nominees of the three companies.

Work began almost immediately on laying out the route of the line and within weeks it was recorded that navvies were already hard at work excavating cuttings at Parsons Rock near Teignmouth. This work highlighted one of the most significant changes to the route originally proposed by Brunel in 1836, in that after leaving Exeter, the route of the new line ran along the right bank of the River Exe to Starcross, before hugging the sea wall past Dawlish to Teignmouth. From there, the most direct route to Plymouth involved steep gradients as the railway skirted the slopes of Dartmoor. This change was announced at the first General Meeting of the South Devon Railway at the end of August 1844. The Chairman Thomas Gill, a local MP and Exeter businessman, reassured shareholders that the changes to the route would more than make up for the costs the company had incurred in new surveys, noted as almost £20,000 in the accounts.

A more momentous announcement was that the Directors had decided to adopt the 'Atmospheric System' of propulsion for their line. Following the passing of the Act in June, they had been approached by Samuel Clegg and Joseph Samuda, engineers who had patented their invention in 1838 and who had claimed their new system would be well suited to the SDR line with its steep gradients and would be a better solution than locomotive power. They also dangled the carrot that the Atmospheric Railway system would be cheaper, an assertion always likely to be of interest to a railway unsure of whether it could raise enough funds to complete a project.

Gill told sceptical shareholders that the board had consulted Brunel who was in favour of the new idea. Although the idea of using air pressure to power vehicles was not new, its employment on railways was as yet untried and very unorthodox. At the time of the shareholders meeting, the only working atmospheric system then in use was the Kingstown to Dalkey line near Dublin, although Brunel had witnessed early experiments in London as early as 1840.[48]. The Duke Street office diary recorded a visit to Dublin on 2 October 1843, and it is likely that Brunel had made other visits to the Irish project in previous years, which allows the possibility that Clegg and Samuda's letter had been prompted by him, although no correspondence survives to confirm this. This hypothesis might well be confirmed by the opening section of a short report to the board written on 19 August recommending the adoption of the atmospheric system. 'The question is not new to me,' Brunel wrote, 'as I have foreseen the possibility of it arising, and have frequently considered it.' Clegg & Samuda's system 'must be cheaper...as it is susceptible of producing much higher speeds than locomotive power', he added.

Before going on to justify his claims, Brunel did concede that his opinion was 'directly opposed to that of Mr Robert Stephenson' who had produced a report on the subject for the Directors of the Chester & Holyhead Railway in April 1844. Stephenson had argued that 'the atmospheric system was not an economical mode of transmitting power' and that it could not maintain higher speeds than the locomotives then in use. While on short lines the system might be 'advantageously applied', on longer railways it was 'inflexible' and not suited to the handling of large amounts of traffic'.[49] Brunel carefully avoided mentioning these conclusions, telling Directors that having considered the subject for several years he thought that 'mechanical difficulties could be overcome' and that as a 'professional man' he held a 'decided opinion that the Atmospheric System has succeeded perfectly as an effective means of working trains whether on long or short lines'. Adding that trains would be faster and with less smoke, cleaner, he ended his report stating that 'I have no hesitation in taking upon myself the full and certain responsibility of recommending the adoption of the Atmospheric System on the South Devon.'[50]

Brunel, Gooch and some of the SDR Directors subsequently visited Dublin to see the Kingstown & Dalkey Railway in action and while the latter may have been impressed, Gooch was not and wrote that Brunel seemed oblivious to the costs and problems apparent, believing that he had so much faith in being able to improve it that he 'shut his eyes to the consequence of failure'.[51] There was, not surprisingly, a protracted discussion at the company General Meeting with questions being asked about speeds and the use of a relatively untried new system over a much larger distance, with one shareholder rightly pointing out that there was quite a difference between the mile-and-three-quarter-long Irish line and the fifty-mile South Devon Railway. When the proposal was put to the vote, it was passed unanimously even though it also meant abandoning the original plans placed before Parliament, which allowed for a doubled-tracked line worked by steam locomotives. In his 19 August report Brunel instead proposed a single-track route with passing places; this was, he reported, on account of the expense, as the cost of the pipes required for the system would be £3,500 per mile, amounting to a total cost of £330,000. These high installation costs would, however, be balanced by reduced running costs since there were no engines to buy, maintain or stable. Brunel estimated that these savings might run to £257,000 initially, with a further saving of £8,000 annually, sums that were no doubt attractive to shareholders.

The working of the atmospheric system has been well explained elsewhere,[52] so it is hoped a brief description here will suffice to summarise the radically different system employed on the SDR to that used elsewhere on Brunel's railways. Cast iron pipes or 'tubes' were laid between the rails and at the front of each train was a carriage attached to a piston inside the pipe; air was pumped out from the tube in front of the piston, allowing atmospheric pressure behind it to push the piston along the tube, moving the train. Stationary steam engines were provided in elegant Italianate engine houses at regular intervals along the route. In theory, the system seemed to have great advantages; there would be little smoke and steam apart from that generated at the engine houses, and since the line was not worked by locomotives, curves could be sharper, and trains could run on the steeper gradients Brunel finally incorporated into the route.[53]

Despite some misgivings, South Devon shareholders seemed convinced by Brunel's arguments. At a subsequent General Meeting, the Chairman made the dubious claim to shareholders that 'the atmospheric system was being adopted in almost every part of the country, and notwithstanding the *Northern Leviathan*[54] in railways who had said it was an "Atmospheric Humbug" the Directors were satisfied' with the system. Reporting to shareholders in February 1845 Brunel gave a list of excuses for lack of progress on the railway, blaming weather and government bureaucracy, but argued that the 'disappointments suffered so far' were drawing to a close, and that it was hoped to open the line to Teignmouth and Newton 'in a short period'.[55] In relation to the atmospheric system, he told them 'delays will unavoidably occur, and in my desire to profit from the experience of the Croydon Railway, I have intentionally and I think wisely postponed mere matters of detail.' The success of the London & Croydon line had confirmed the suitability of the system, he concluded.[56].

Brunel's statement was disingenuous, and his answers when questioned by a Select Committee on Atmospheric Railways in April 1845 seem to suggest that many of the 'mere details' he had noted were still to be worked out. Asked by the Chairman if he had an idea of the manner in which the line would be constructed, Brunel gave the astonishing reply 'To a certain extent but not entirely' – the new railway required 'many new contrivances', details of which he added, he had not yet been completed to his satisfaction.[57] Despite the physical completion of the line between Exeter and Teignmouth by the spring of 1846, progress on the installation of the atmospheric system was painfully slow, and the line had to be operated initially by steam locomotives hired from the GWR. Brunel reported to shareholders that this could be done with 'perfect facility' and would not interfere with the installation of the atmospheric equipment. The first trains between Exeter and Teignmouth ran on 30 May 1846 with the section to Newton Abbot opened on 31 December. The Chairman was forced to continue making reassuring announcements at shareholders meetings as work dragged on to construct the engine houses, install stationary steam engines, and lay the pipes as the permanent way was made ready. Brunel had originally intended to use three sizes of pipe; thirteen-inch diameter for the easily graded Exeter-Newton section, twenty-two inch for the steep gradients west of Newton and fifteen inch for the remainder of the line to Plymouth. More than 4,000 tons of 13-inch pipe had been manufactured before the engineer changed his mind and opted for 15-inch diameter instead, a change that cost the SDR a further £31,000 it could ill afford.

Finally, the atmospheric pipes began to be installed in December 1845, but by April the following year only seven miles had been laid. In September Brunel reported 'very slow progress' to shareholders noting that while pipes had been laid almost all the way to Newton Abbot, the pumping engines were not ready. The delays were regrettable he added, once again using the excuse of gaining more experience from the working of the Croydon line[58] and reassuring the proprietors that the result would repay them 'for the anxiety of waiting'.[59] The Directors were not only concerned at lack of progress but also at Brunel's absence from board meetings, perhaps as much a reflection of embarrassment on his part as pressure of work elsewhere. It seems that Brunel had been asked to appoint a deputy by the board, but in typical fashion resolutely refused to do so. In July 1846 he was therefore formally asked to appear before them and report on progress, and a sub-committee was then set up to review matters more closely going forward.[60]

The first experimental trains ran using the atmospheric system in February 1847 over a 20-mile stretch of line between Exeter and Newton Abbot, but more than a year later the 'Atmospheric Caper' as it became known was soon proving to be a liability to the SDR. Brunel's position was not improved by the news in May 1847 of the closure of the Croydon Railway, clearly a blow. The engineer was not present at a board meeting held shortly after and it was the ever-loyal Chairman Gill who had to report that Brunel had given 'explanations of a satisfactory nature'. It was reported however that Joseph Samuda was in attendance at the board meeting, and he seems to have taken an active role in supervising the work. A *Railway Times* article also seems to support the greater role played by Samuda, noting that two pumping engines had recently been commissioned, 'under the immediate superintendence of Mr Samuda himself, who is frequently on the spot'.

More difficult questions were asked at the next General Meeting when it was announced that a public service using atmospheric power was imminent, but a timetabled service did run from

September 1847 and over the next few months locomotive-hauled services were gradually phased out so that by February 1848 all goods and passenger services between Exeter and Newton were run using atmospheric power. Trains had reached speeds of between forty and fifty mph during tests and it seemed that the 'Atmospheric Caper' might finally be successful. Passengers were happy with the lack of smoke and cinders and one reporter remarked on the smooth ride, recording that 'you can write while the speed is 50mph or more,' a sentiment Brunel would have surely approved of. However, this positive news masked ongoing and increasing operational difficulties which turned out to be both practical and financial. These largely involved the difficulties experienced in maintaining an air-tight seal of the cylinder piston. The leather valve flap fixed to the top of the pipe to maintain an airtight seal suffered badly in the salty atmosphere of the Devon coast and extra staff had to be employed by the SDR to continuously grease the leather. The leather was also affected by frost and the effects of sun and wind, tearing and causing leaks and with air leaking from pipes, pumps had to work harder, which increased the amount of coal consumed. Brunel continued to provide a reassuring narrative for shareholders telling them in February 1848 that he was 'in a fair way shortly of overcoming the mechanical defects and bringing the whole apparatus into regular and efficient practical working'. Brunel blamed some difficulties on the lack of a telegraph system that could allow engine houses to communicate, neglecting to mention that he had commissioned this a year earlier but had not managed its installation properly.

As spring turned to summer, the leather valves continued to deteriorate and more than two miles of it was replaced at a cost of more than £1,000. When it became apparent that maintenance costs were eating into operating profits and that the atmospheric system was more costly to run than locomotive power, a special committee was convened to investigate and Brunel was asked to prepare a report on the situation. After failing to appear in Devon personally, he was visited by a deputation at Duke Street that included Gill, GWR Chairman Charles Russell and Buller, Chairman of the Bristol & Exeter. There is no record of what was said at the meeting, but the gloomy verdict Brunel delivered in his formal report of 19 August was a contrast to the bullish optimism displayed in 1844. No further improvements to the atmospheric system could be made without the complete replacement of the valve, which he described as 'not being in good order' and more money would also be required to increase the power of the pumping engines, which his original specification had left underpowered. In a final humiliation, he concluded that given all that had transpired so far, he could not recommend the extension of the system beyond Newton Abbot and on to Plymouth. Samuda had offered to make improvements to the valve at a cost of £210 per mile, but the board were in no mood to spend more money and insisted they would only agree to this if Samuda would also take on the entire cost of working the line. Unless such a guarantee was provided by Samuda they agreed, they would suspend the use of the atmospheric system on 9 September 1848.

The final indignity was a shareholder meeting on 29 August, and this time Brunel could not hide away in London, instead having to face a barrage of criticism in public, although SDR Directors were also roundly condemned by investors who had lost considerable sums. The engineer was described as reckless and extravagant and Brunel's son later recorded with some understatement the platitude that his father had been 'much censured' over the 'Atmospheric Caper' but that he had recommended the use of the system 'with a simple and self-sacrificing disregard of every consideration except that which was always paramount with him, the interests of those by whom he was employed'.[61] Although it emerged in the meeting that Brunel had lost £20,000 of his own money in the venture, and that he had also agreed to waive his fee for working on the railway, Directors and shareholders and the eighty staff who had lost their jobs when the Atmospheric system was abandoned must have had a different perspective.

Although locomotives took over the running of trains, the equipment and pumping engines remained in situ and debates about the atmospheric system rumbled on. After a marathon eight-hour meeting in January 1849 shareholders finally voted to abandon it completely, leaving the South Devon Railway with a thumping loss of £433,991. The company was able to recoup more than £80,000 by selling off pumping engines, pipes and other equipment including the engine houses, but the deficit along with the expense of having to re-equip the line to run with conventional steam traction was keenly felt by the company for years. The curves and gradients on the line engineered by Brunel would also provide challenges for steam locomotives, with banking engines or double-heading required over the hills beyond Newton Abbot. In addition, the decision to route the railway along the sea wall would also provide further problems for the South Devon and later the GWR, British Railways and Network Rail today. When describing the route, Brunel boldly stated that water would only break over the sea wall 'in extraordinary circumstances of high spring tides and heavy gales', arguing that 'a railway is not a work liable to any considerable damage by the beating of water over it.' As early as October 1846 gales had already severely damaged the railway near Dawlish and Brunel was personally supervising the strengthening of the sea defences, *The Times* reporting that between eighty and 150 men were employed building a new limestone wall backfilled with sandstone. The Directors, though 'visited with loss and vexation' could be considered 'more free from responsibility to the shareholders and public than the engineer' the article noted, reporting that this stretch of coast was one of the 'wildest', a fact not disputed by the engineers who have had to repair serious breaches in the sea wall in the years since.[62]

The whole atmospheric saga was a great embarrassment to Brunel and a serious blow to his reputation, his friend Gooch recording in his diary that he could not understand 'how Mr Brunel was as misled as he was'. Not only had he once again been seduced by the prospects of a new technology without thinking through the practical consequences, he had also provided yet another example of his over-optimistic cost estimates for his railways. His original estimate for equipping the whole line from Exeter to Plymouth was only £190,000, a sum more than exceeded in the twenty-mile stretch built to Samuda's specifications. Brunel's absences during the construction of the railway which had so exasperated Directors also highlighted another less welcome trait that remained a feature of his career, his excessive workload and inability to delegate to assistants, which meant that despite working longer and harder than most, even he could not keep up with the demands placed on him.

The abandonment of the atmospheric system on the South Devon was one of the reasons for the demise of the Portbury Pier & Railway Company, a Bristol project Brunel had been involved with on and off for more than fifteen years. As early as 1832 he had suggested the construction of a pier at Portishead that could be used by mail packets to avoid a slow trip down the river to the docks in the city. Following the first voyage of the SS *Great Western* in 1838, the Great Western Steamship Company made representations to the Bristol Dock Company to reduce dock dues, but were rebuffed, leading to the ship being moved away from the city to Liverpool. The intransigent attitude of the dock company and its reluctance to invest in facilities such as lock gates led to calls to build a new dock downstream in an attempt to prevent business being driven away from the city.

Attempts were made to progress the idea in 1840 and 1841, but at a meeting held in April 1845 a provisional committee of the Portbury Pier and Railway Company was created and resolved that it was 'essential for the maintenance and advancement of the traffic of the Port of Bristol' that accommodation was made to enable passengers and goods to be landed and embarked 'at any state of the tide' and that this could be achieved by the construction of a floating pier at Portbury, designed by Brunel, and a railway of about six miles that would link it with the Bristol & Exeter and Great Western railways. The report produced by the committee noted that the line was to be single 'and worked upon the atmospheric principle, for the application of which it is from its gradients and length peculiarly adapted'.[63] Sketches by Brunel of the proposed pier at Portishead from as 1839 survive, but little detail about the railway has yet been discovered. By 1847 the 'depressed state of the money market' meant that the company did not yet have enough capital to begin work. Directors debated whether to build the pier without the railway, and Director William Miles told his colleagues that he supposed 'Mr Brunel had been too much occupied to make a report, though he had been pressed over and over again to do so.'[64] It is likely that Brunel's mind was at that point preoccupied with events in Devon and the failure of the atmospheric system there, along with a collapse in investor confidence as the 'Railway Mania' came to an abrupt halt and the fortunes of the Portbury Pier & Railway Company were to follow a similar trajectory. Work was suspended in 1848, and it was not until 1863 that another attempt was made to build a railway to Portbury and Portishead, this time without atmospheric traction as a motive power.

8

Onwards

It might have appeared that the Gauge Commission had drawn a line under the whole issue of the broad gauge and that its fate was finally sealed. Although the question of a unified national gauge of 4 ft 8 ½ in had been decided, the concessions subsequently made to the GWR meant Brunel would have more than enough work to do in the coming years as engineer to a significant number of railways which had gained Parliamentary assent before the Commission ruling. A Board of Trade minute published in June 1846 produced in advance of the final Gauge Act had provided some further insight into their more pragmatic view of the implications of the Commission's findings. The Board did not feel justified in recommending that all existing broad gauge lines should be narrowed using public money at 'vast expense', or that any broad gauge companies should be compelled to incur those costs bearing in mind they had embarked on construction of their lines with the sanction of Parliament. The best course, given that 'uniformity of gauge' had now been established, they concluded, was a 'settlement of gauge' on the railways that already had Acts of Parliament and regulations to prevent the increase and further extension of 'an evil they could not altogether remedy'.[1]

In practical terms this ruling and the subsequent Act of Parliament[2] meant that, apart from the railways explicitly listed in the legislation, no further broad gauge railways would be allowed. Even before the Act had been passed, a sign of things to come had been the agreement by the GWR board in July 1846 that an additional 'narrow' gauge rail would be laid on the contentious Oxford & Rugby line and also on a forty-mile stretch of line between Oxford and Basingstoke that included the Berks & Hants Railway. The latter was the result of a concession by the Board of Trade not to sanction a rival narrow gauge scheme, the Manchester & Southampton Railway then being considered by Parliament. So it was that the GWR retained an important broad gauge route, but by conceding that a third rail would be added laid the foundations for the eventual demise of Brunel's innovation. The spread of what became known as 'mixed gauge' lines would be inevitable, but for now Brunel had some other challenging railway projects to engineer and there would also be further struggles to be fought as a rear-guard action. The consequences of the Battle of the Gauges lingered on until the mid-1850s.

The Oxford, Worcester & Wolverhampton Railway had been one of the original schemes which had precipitated the Gauge Commission debate and over which the whole battle had been fought. It had been named as one of the railways which could be built to the broad gauge as originally proposed, but Brunel soon became caught up in a dispute between the OW&WR and the Great Western Railway, the latter having previously provided important financial and parliamentary support. MacDermot, the Great Western Railway's own historian, later argued that the broad gauge had been 'stabbed in the back by the mutiny of the Oxford, Worcester & Wolverhampton Railway' and took a partisan view on what was a complicated story. The ninety-mile railway began at Oxford and Brunel in planning the route faced some significant engineering challenges. After following the course of the River Evenlode for almost thirty miles, the line then climbed through the Cotswolds by way of Charlbury and Kingham before dropping down from a summit near Moreton-in-Marsh to cross the more easily graded Vale of Evesham. The line then pushed north through Worcester and on to Wolverhampton. As well as numerous bridges and viaducts along the route, the OW&WR also featured three tunnels: one at Mickelton near Chipping Campden, and two others at Wolverhampton and Dudley.

The first General Meeting of the company held on 14 October 1845 did not get off to the most auspicious start when irritated shareholders were kept waiting for almost an hour by the new Chairman Francis Rufford; his Directors and Charles Saunders, the Chairman said, had been detained by 'a deputation and other matters'. Rufford reported more positively that the parliamentary struggle had been 'arduous' but their success was 'a matter of universal rejoicing throughout the district'. Brunel does not appear to have been in attendance that day, but the Chairman was also able to announce that engineers had been working hard to complete final detailed surveys and 'boring the ground to ascertain the nature of the soil'. Particular care was being taken where the line passed through manufacturing and mining areas he added. Work was also proceeding 'on procuring the land and commencing the works'.[3]

What Rufford did not mention was that it had become apparent shortly after the passing of the OW&WR Act in August 1845 that Brunel's original estimate of £1,500,000 would not be enough to complete the railway. Rufford wrote to the Directors of the Great Western Railway on 22 November requesting that the existing arrangement between the two companies should be modified and asking them to increase the amount they would guarantee 'to cover the cost of the line as originally contemplated'.[4] Rufford cited increases in preliminary expenses incurred during the parliamentary contest, large sums paid to landowners, the cost of making the line between Abbotswood and Wolverhampton double track as insisted on by MPs and the cost of further branch lines, also a result of discussions in Parliament. Rufford continued that 'Mr Brunel cannot at present re-estimate the cost of construction and additional costs' but asked for confirmation that an arrangement of this type might be possible. The GWR Chairman Russell replied reassuring Rufford that his grounds for such a request were reasonable, and that his company would 'go into the subject in the best spirit'. Brunel did not complete the revised estimate until February 1846 when he reported that the cost of the line would now be £2,249,914, almost one million pounds more than his original estimate. This included an allowance of £91,548 covering parliamentary expenses, additional surveys and other expenses as well as interest charges of £150,000. While some additional expenditure could not be attributed to Brunel directly, he had clearly underestimated construction costs yet again.

The GWR board agreed to extend their guarantee to cover a sum up to the £2,500,000 required by Brunel and the OW&WR to finish the line, in the process raising the rate of interest payable from 3½ to 4 per cent. When the matter was put to the GWR shareholders a few days later, the wording of the resolution had been subtly altered so that it read that the Great Western would only guarantee a 'such a sum as shall appear to them necessary for the completion of the said railway and works'.[5] When this more nebulous proposition was discussed by the OW&WR shareholders some weeks later there was uproar, as many had assumed that the Great Western would back them whatever the cost. This change marked the beginning of the unravelling of the relationship between the companies and it is difficult to judge whether it was a misunderstanding or something more deliberate. In truth, the Worcester company had little option but to accept either the original offer of guarantee from the GWR or its modified deal. Writing to Lord Redesdale, a critic of the arrangement and the landowner affected by the passage of the railway across his land in Moreton-In-Marsh, Rufford noted that the lease with the GWR had been the 'foundation of our Company' and that the original railway proposal had failed to attract enough subscribers until the guarantee from the Great Western Railway had been advertised. Redesdale was also critical of suggestions that the Worcester company should abandon the broad gauge or sell to the GWR. He recommended that the OW&WR should 'keep what we have got and not be bamboozled or trampled upon'.[6]

With the backdrop of turmoil over the company's financial arrangements it was not surprising that the process of starting construction of the railway took longer than Brunel hoped. By the spring of 1846 it was reported that there had been 'much dissatisfaction' in the neighbourhood of Worcester at delays in starting work on the line and that there had been suspicions 'freely expressed' about the viability of the company. The same report recorded that test borings had taken place near Dudley on the land of Lord Ward, but his agents had insisted that purchase money be paid before the property was disturbed.[7] Given the disruption caused by the Gauge Commission and its aftermath, the caution of the OW&WR Directors was not surprising, but work to build the railway finally began in the summer of 1846 and Brunel was able to report that excavations had begun on the tunnels soon after. Another correspondent from *The Railway Times* reported a 'hasty run' over the line in October and wrote that great progress had already been made: 'The heavy parts of the railway are all in the hands of the "navvies" and the contractors are apparently vigorously doing their duty.' At the north end construction work was so advanced that it was hoped some sections could be opened by the following year. The correspondent also reported that there had been 'silly rumours' heard in Worcester that the OW&WR was to be purchased by the London & Birmingham or that the GWR had sold its shares in the railway. These were 'ridiculous and unfounded' and the company was in 'a most flourishing state'. When finished the line would give a 'respectable dividend, over and above the 4 per cent guarantee of the Great Western Railway', the report concluded.[8]

The optimism portrayed in the report was shortlived. The state of the national economy, badly effected by the famine in Ireland, revolutions in Europe in 1848 and the collapse of the railway stock market as the 'Railway Mania' began to burn itself out, meant that additional capital became hard to find. By July 1848 the Oxford, Worcester & Wolverhampton had run out of money and with no section of line yet open there was nothing to show for the capital already invested. GWR Directors refused to provide any further capital and Brunel was clearly unhappy at the state of affairs. Following discussions between the two companies it was agreed that it might be expedient for any work to be concentrated on the Oxford to Worcester section only; Brunel was asked to report and wrote to Rufford on 8 September 1848. His frustration at the situation was clear, agreeing that it would be better to concentrate the funds available for the 'earlier completion of some portion (of the line) which can be rendered profitable'. He had he said 'felt compelled to urge this very frequently and I am still without instructions', noting that no arrangements had been agreed to reduce expenditure on any one section of line. 'Let me beg you to give me some authority to act,' he wrote, to prevent further waste of money on work on sections of line that could be postponed. 'My conviction is that you have no alternative but to suspend all expense beyond Stourbridge and devote every sixpence you can raise to the south end of the line.'[9]

Brunel's recommendation was endorsed by the board but by early 1849 the position was little better and a committee of investigation appointed by OW&WR shareholders determined that a further £1,500,000 would be required to complete the line. Some idea of the pressures faced by both Brunel and the company come in an entry in his journal for 1 May 1849. Reporting on a meeting with the Directors he wrote that he had told them that he had 'kept down payments to contractors in every possible way fortnight after fortnight' and that the time was fast approaching when 'we must either provide more money for carrying on the work or must arrange with the contractors to wind up.' He gloomily concluded: 'The Directors said – and I felt that they could not say otherwise – that at present they could not determine upon any particular plan of proceeding and that things must stand over a little longer.'[10]

The GWR board, its patience perhaps at an end, refused to assist and the 'mutiny' gained strength. At a shareholders meeting in January 1851 it was recorded that since negotiations with 'other connected companies' had failed it was now time to put the company on a proper footing as 'an independent line'. It was also announced that contractors Peto and Betts would take over the Oxford to Worcester and Wolverhampton to Tipton sections of line, promising to complete this work within eighteen months.[11] Another sign of this new-found 'independence' was to abandon the broad gauge and when the Company announced it would only lay narrow gauge track south of Abbotswood, the GWR served it with an injunction in Chancery stating the Worcester Company was in breach of its agreement and its Act of Parliament.

There is not space here to tell the full story of the long-running dispute between the two railways, but Brunel, caught in an invidious position as engineer to both the OW&WR and the Great Western Railway, resigned from his role for the Worcester company in March 1852 'except to complete and settle certain pending contracts and accounts'.[12] The GWR board had taken legal advice regarding Brunel's position and had reported that unless both the OW&WR and the GWR could agree that he was neutral as engineer of both companies, 'his position will prejudice that case brought by the GWR against the OW&WR to restrain the OW&WR from laying the standard gauge rail'.[13] Brunel had also faced hostile criticism from OW&WR shareholders and Directors about his apparently excessive use of company funds. His successor was John Fowler, later to become engineer of the Metropolitan Railway. A GWR account noted, however, that in December Brunel had been called on to report on the state of the OW&WR line 'so far as under the circumstances he was able'. He stated that over a considerable distance of the railway there were no broad gauge rails laid and where it was in place the formation was 'highly dangerous' and 'not constructed or being constructed to his satisfaction'.[14]

The Oxford, Worcester & Wolverhampton Railway was eventually completed in stages beginning with the Droitwich-Stourbridge section, which opened in May 1852. The line had been completed as far as Evesham by the spring of the following year, but the opening of the final 'Cotswold' section of the railway was delayed when the Board of Trade inspector refused to pass the line since sections of the broad gauge track were still unfinished. Trains finally ran between Dudley and Oxford in June 1853. One final section of the railway remained unfinished. The last five miles into a new station at Wolverhampton Low Level were not completed until July 1854, much of the delay due to ongoing arguments between the OW&WR and the GWR.

As a result of the disputes between the companies, Brunel does not appear to have taken part in any of the festivities surrounding the completion of the Oxford, Worcester & Wolverhampton Railway, something that might have been a disappointment given that he had been so involved for much of the decade-long struggle to complete the line. The design of the railway had all the hallmarks of a Brunel line; as well as his baulk road track largely in a mixed gauge formation, the railway also featured some of his standardised single-storey station buildings. Since funds were short, most were basic wooden structures with small canopies to shelter waiting passengers.[15] These were a far cry from the grander structures seen on the Great Western Railway and elsewhere, although larger chalet-style buildings were provided at some locations similar to other smaller Brunel stations seen on the Bristol & Exeter and South Devon lines. Charlbury is the only surviving example of this design, an elegant wooden Italianate timber building with hipped roof forming an integrated canopy to protect passengers from the elements.[16]

The OW&WR also featured a large number of timber bridges and viaducts, reflecting the modest budgets within which Brunel was working. There were more than sixty on the main line, many concentrated north of Worcester and most built between 1852 and 1854.[17] All were built to accommodate a double line of broad gauge track and there were a number of imposing structures; most impressive was the Hoo Brook viaduct south of Kidderminster, which carried the railway across the valley on twenty-two trestle spans seventy-six feet high.

Of the three tunnels on the railway, it was the construction of the tunnel at Mickleton near Chipping Campden which was most notable and not just for its engineering features. The original contract to build the tunnel had been awarded in 1846 to William Williams & Robert Mudge-Marchant, Brunel's second cousin, but relations between engineer and contractor had deteriorated over payments, the situation exacerbated by work being suspended when the finances of the OW&WR had become difficult. In December 1850 Brunel told Williams that he would recommend the company 'pay a fair amount for the cost of maintenance of the works while work was at a standstill' but added he was 'not speaking on the part of the company and cannot bind them, but merely state what I shall consider fair'. A few weeks later Brunel wanted to know 'immediately' what steps the contractors were taking to proceed with the works with particular reference to the supply of bricks, 'as there appears to me to be a great want of good bricks in stock – a large proportion of those now there being unfit for use'. In early January Williams wrote to Brunel arguing that their claim for maintenance and compensation had still not been addressed and that Williams and Marchant were not 'persuaded that the Directors when they really want us to proceed with dispatch will not withhold our resources'. Brunel replied that Directors had given him the power to settle the dispute and report back and that if he had to do so unfavourably, 'I fear that consequences will be disadvantageous to you' and that it was 'five weeks since I wrote to you and since you ought to have proceeded with the works'.[18] Some kind of compromise appeared to have been agreed but the dispute rumbled on and by June 1851 Williams and Mudge-Marchant downed tools again, claiming that they were owed £34,000 by the OW&WR.

The Directors told Brunel to evict the contractors from the site and seize equipment to compensate them for the money they thought Williams and Rudge-Marchant owed them. On the evening of 17 July 1851 Brunel arrived at the tunnel in his 'Flying Hearse' carriage with hundreds of navvies who had marched from

1. A lithographic image of J. C. Horsley's portrait of Isambard Kingdom Brunel. (STEAM: Museum of the Great Western Railway)

2. A nineteenth-century illustration of railway surveyors at work. (Author's collection)

3. Brunel's Duke Street office, probably photographed after his death in 1859. (Author's collection)

Right: 4. A page from Brunel's notebook, written and sketched as he travelled on the Liverpool & Manchester Railway in 1831. (By courtesy of the Brunel Institute – a collaboration of the SS Great Britain Trust and University of Bristol)

Below: 5. The first public meeting of the GWR held in 1833, reimagined by the GWR in 1835 for a publicity film *Romance of a Railway*. (Author's collection)

ANNO QUINTO & SEXTO

GULIELMI IV. REGIS.

****

Cap. cvii.

An Act for making a Railway from *Bristol* to join the *London* and *Birmingham* Railway near *London,* to be called "The Great Western Railway," with Branches therefrom to the Towns of *Bradford* and *Trowbridge* in the County of *Wilts.*

[31st *August* 1835.]

WHEREAS the making a Railway from *Bristol* to join the *London* and *Birmingham* Railway near *London,* and also Branch Railways therefrom to *Trowbridge* and *Bradford* in the County of *Wilts,* would be of great public Advantage, not only by opening an additional, certain, and expeditious Communication between the Cities and Towns aforesaid and the several intermediate and adjacent Places, but also by improving the existing Communication between the Metropolis and the Western Districts of *England,* the South of *Ireland,* and *Wales:* And whereas the King's most Excellent Majesty in right of His Crown is entitled to certain Lands upon the Line of the proposed Railway: And whereas the several Persons herein-after named are willing at their own Expence to carry into execution the before-mentioned Undertaking; but the same cannot be effected without the Authority of Parliament: May it therefore please Your Majesty that it may be enacted; and be it enacted by the King's most Excellent Majesty, by and with the Advice and Consent of the Lords Spiritual and Temporal, and Com-

[*Local.*] 35 O mons,

6. The Act of Parliament authorising the Great Western Railway on 31 August 1835. (SS Great Britain Trust)

7. The Wharncliffe Viaduct, the first major structure to be built on Brunel's Great Western Railway. (SS Great Britain Trust)

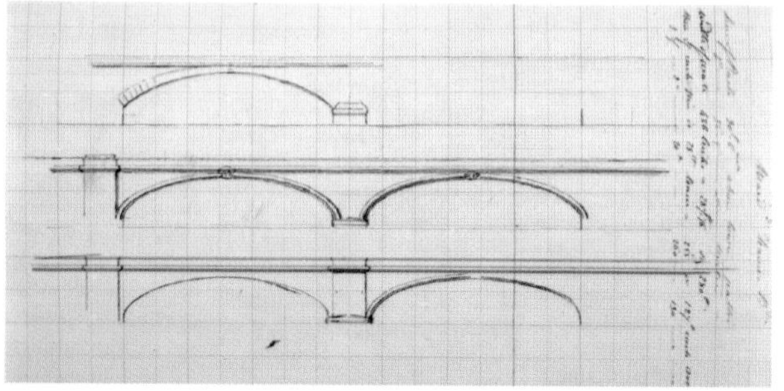

8. One of Brunel's original sketch designs for the Maidenhead Bridge. (By courtesy of the Brunel Institute – a collaboration of the SS Great Britain Trust and University of Bristol)

Above: 9. The imposing west portal of Box Tunnel. (SS Great Britain Trust)

Right: 10. The interior of Box Tunnel showing the unlined interior. (SS Great Britain Trust)

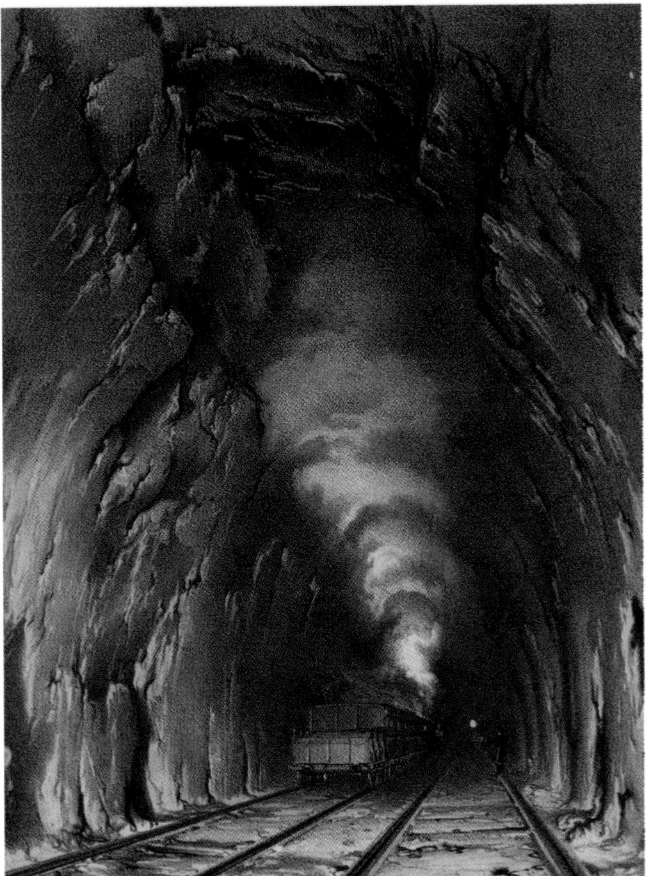

11. Two pages from Brunel's sketchbooks showing designs for lamp posts at Bath Station. (By courtesy of the Brunel Institute – a collaboration of the SS Great Britain Trust and University of Bristol)

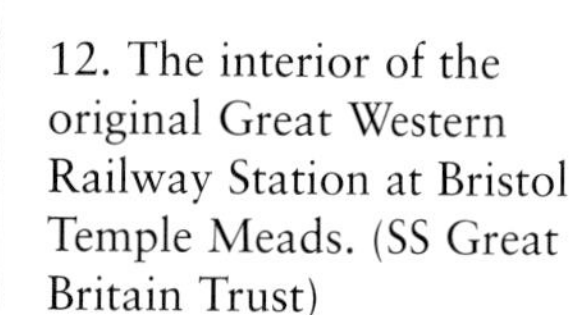

12. The interior of the original Great Western Railway Station at Bristol Temple Meads. (SS Great Britain Trust)

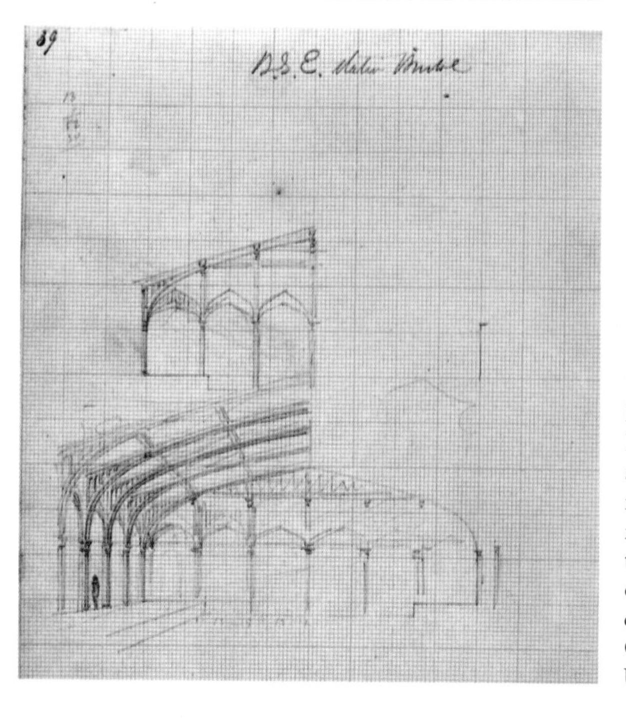

13. A sketch by Brunel of his original design for the Bristol & Exeter terminus at Temple Meads; it shows many similarities to the roof of the 1840 GWR train shed. (By courtesy of the Brunel Institute – a collaboration of the SS Great Britain Trust and University of Bristol)

14. Brunel's broad gauge 'Baulk Road' track as recreated at the Great Western Society's Didcot Railway Centre. Strictly speaking the image shows mixed gauge track, which allowed both broad and narrow gauge trains to run. (Author's collection)

15. Brunel's Transatlantic dream: London to Bristol by GWR and on to New York on his steamship the SS *Great Western*. (Author's collection)

Above left: 16. The infamous Refreshment Rooms at Swindon seen in Measom's 1852 *Guide to the GWR*. (Author's collection)

Below left: 17. The Engine House at Swindon Works in 1846 as portrayed by J. C. Bourne in 1846. (SS Great Britain Trust)

Below: 18. The original broad gauge engine shed at Swindon, which may well have been moved from Hay Lane to Swindon before the works opened in 1843. This photograph shows the old building prior to demolition in 1930. (STEAM: Museum of the Great Western Railway)

Right: 19. Daniel Gooch, the first Locomotive Superintendent of the Great Western Railway. He stands close to a model of one of his Firefly class locomotives. (STEAM: Museum of the Great Western Railway)

Below: 20. The Break of Gauge at Gloucester as portrayed by the *Illustrated London News* in 1846. (STEAM: Museum of the Great Western Railway)

Left: 21. An 1845 map of Brunel's broad gauge network and the Breaks of Gauge. (Author's collection)

Below: 22. Brimscombe Station on the Cheltenham & Great Western Union Railway, opened in 1841 but pictured in 1954, a good example of one of Brunel's smaller, chalet-style stations. (STEAM: Museum of the Great Western Railway)

23. A rare and very early photograph of Brunel's Chepstow Bridge over the River Wye taken following the raising of the second tube, which is still in the process of being lifted to its final position. (SS Great Britain Trust)

24. The Landore Viaduct on the South Wales Railway, one of the largest timber viaducts designed by Brunel. (Author's collection)

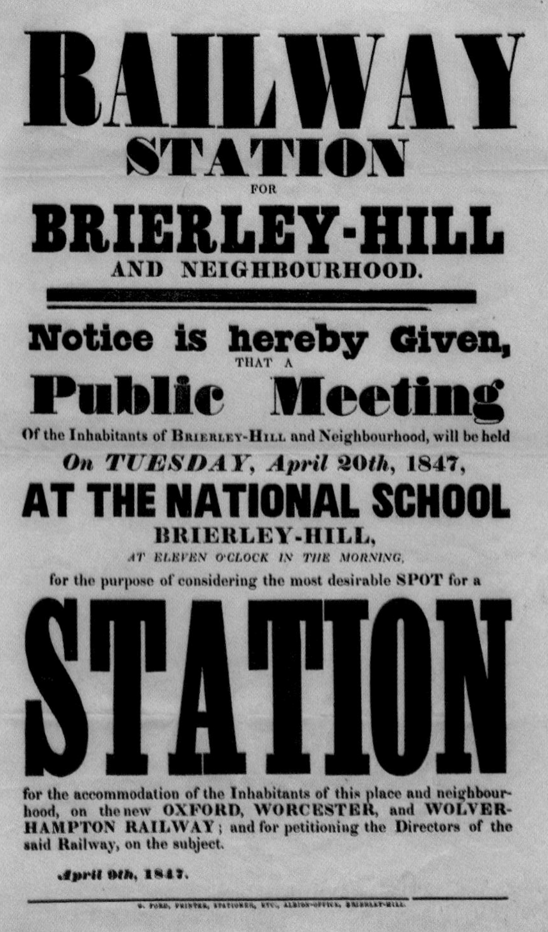

Above: 25. Celebrations at Swansea marking the opening of the South Wales Railway from Chepstow in 1850. Daniel Gooch was on the footplate of one of the locomotives hauling the first train. (Author's collection)

Left: 26. A rare handbill for the Oxford, Worcester & Wolverhampton Railway. (SS Great Britain Trust)

27. Charlbury station on the first stretch of the Oxford, Worcester & Wolverhampton Railway between Oxford and Worcester. The line has been restored to a double track formation since the picture was taken. (Author's collection)

28. Slade Viaduct, built for the South Devon Railway, seen in 1892 as a replacement masonry viaduct is constructed next to it. (STEAM: Museum of the Great Western Railway)

Top: 29. Newton Abbot atmospheric pumping house and station. There are no signs of any trains in the picture. (STEAM: Museum of the Great Western Railway)

Above: 30. A somewhat ghostly stereoscopic view of Brunel's new Paddington Station in the years after its completion in 1854. (SS Great Britain Trust)

Left: 31. One of Brunel's sketches for of Paddington Station, completed at Duke Street on 20 December 1850. (By courtesy of the Brunel Institute – a collaboration of the SS Great Britain Trust and University of Bristol)

Above: 32. The Royal Albert Bridge showing the eastern bridge truss in the process of being slowly inched up into position as the masonry of the bridge pier is completed. (SS Great Britain Trust)

Right: 33. The distinctive signature of Isambard Kingdom Brunel on a working document written in 1847, one of thousands of individual letters, memoranda, plans and drawings he must have signed during his career as an engineer. (Author's collection)

Mr Barlow and Mr Mills And Thomas H A Poynder Esq, by which the Company agree to purchase certain portions of Mr Poynder's Land that will be required for the Railway for £1600 and to construct the Line through his Estate as therein mentioned

7th April 1847 – Between Charles Russell Esq F. P. Barlow and E W Mills Esquires And His Grace The Duke of Northumberland for the construction of The Railway through His Grace's Estate in the manner therein mentioned

15 April 1847 – Between The Great Western Railway Company and Thomas Wilmot Esq – by which the Company agree in the event of their obtaining their Bill to purchase certain premises at Egham Hithe for £1500

IKBrunel

Above: 34. The end of Brunel's broad gauge dream: the final abolition of the 7-foot gauge in 1892 at Saltash. Thousands of navvies worked over the weekend of 21 and 22 May to complete the conversion of the last sections of line. (Author's collection)

Left: 35. A world-weary Isambard Kingdom Brunel pictured in the last years of his life. (SS Great Britain Trust)

Banbury with him, but the contractors had been warned of his arrival and the local magistrates had been called out. They warned Brunel that he was causing a breach of the peace and, after some brawling and the reading of the Riot Act, Brunel called off his navvies and retreated. Matters did not improve the following day as thousands of other workmen drawn from other nearby Brunel railway projects descended on Mickelton. Fighting broke out, but the appearance of magistrates again and a warning to Brunel that he was responsible for a breach of the peace eventually led to a suspension of hostilities. The three thousand men were dispersed, and Williams and Rudge-Marchant finally conceded defeat with the promise of an arbitrated settlement. This took place in October 1852 when the OW&WR decision to dismiss the contractors was upheld while awarding them compensation for the loss of their contract. It was too late for Rudge-Marchant however, as he had been declared bankrupt in November 1851. Although most injuries suffered in the fighting were relatively minor, in inciting the disturbances Brunel had acted with a recklessness that would have merited arrest and conviction for most people of the time. His undoubted celebrity and influence saved him from this outcome.

The construction of railways in Wales would take up much of Brunel's time in the years after 1846, in particular the South Wales Railway, identified in the 1846 Gauge Act as one of the lines that could still be built to his broad gauge. With the completion of the Taff Vale Railway in 1840, Brunel had turned his attention to railways away from the principality but from 1844 to 1859 he would engineer a number of other lines including the Vale of Neath and design docks such as at Briton Ferry. A small footnote in his Welsh interests provides an insight into a broad gauge railway that might have been. In 1840 Government commissioners had ruled that Holyhead should be the main port of call for trade with Ireland. This led to the promotion of the Chester & Holyhead Railway engineered by Robert Stephenson, which opened to the port in 1848. There had also been proposals to build a new harbour at Porth Dynllaen[19] on the north coast of the Lleyn Peninsula; suggestions were made that this port could be linked to the capital by a railway which crossed Wales, providing an alternative to the coastal route. The engineer Charles Vignoles, who had already been involved in discussions about Anglo-Irish traffic, approached

Brunel and the GWR and they jointly put together proposals for a broad gauge line from Worcester to the port. The Worcester & Porth Dynllaen Railway Company was then incorporated with a capital of £2,700,000 and a proposed route which would run from Oxford via Worcester, Tenbury, Ludlow, Newtown, Dolgelly and Barmouth to Porth Dynllaen.

Early publicity optimistically stated the 256-mile route would only have seven and a half miles of tunnels, boasting that this was less than the total for the London & Birmingham Railway. With the adoption of the broad gauge a speed of 40 miles per hour could be maintained, providing not only 'national advantages' but also 'access to mineral and agricultural riches that abound, and which at present lie dormant for want of efficient means of transit'.[20] The proposed line was mentioned by Francis Rufford, Chairman of the OW&WR at its first General Meeting in October 1845 and he told shareholders that surveys had proved 'highly satisfactory'. Brunel's proposals had not, however, been greeted with universal enthusiasm as the real scale of earthworks required and the steep gradients likely to be encountered became apparent. One newspaper report argued that it was 'to be lamented that Brunel had no more consideration for the profession to which he belongs' and that the railway 'would be a standing joke against him until the end of his life'.[21] A broad gauge main line driven through the Cambrian Mountains would have been an enormous and expensive undertaking with many engineering challenges and while it could have been another potential professional disaster for Brunel, it remains an intriguing prospect none the less. The optimism displayed by the promoters, typical of the 'Railway Mania' period, was ultimately to be to no avail as the bill for the new railway only got as far as its second reading in Parliament before the Gauge Commission report and Act meant the possibility of a broad gauge line to Holyhead was dropped.

Although Brunel's ambition to build a broad gauge main line to North Wales did not come to fruition, he was able to press on with plans further south. By 1845 the embryonic railway network in Britain was taking shape with more than 2,200 miles of track in place, much of which had been added since 1840. Brunel's railways had linked the capital with Bristol and had now reached as far south west as Exeter, but apart from his Taff Vale Railway linking

Cardiff and Merthyr, Wales remained largely undeveloped as far as main line railways were concerned. Brunel had been appointed as engineer of the Gloucester & South Wales Railway in 1836, an early scheme that was an attempt to extend the GWR into South Wales and spent time considering and surveying a route in the following months, despite his other responsibilities on the GWR. His office diary for June 1836 noted that on 23 and 24 June he had been working between Chepstow and Newport, and on 26 June he was 'engaged on the line between Cardiff and Swansea' ending his stay 'travelling all night' to Cheltenham where he had a meeting the following day.[22] By 1837, however, the proposed railway had been abandoned since not enough subscriptions from shareholders had been forthcoming.

Brunel had estimated that it might cost just under a million pounds to build the railway from Gloucester to Cardiff, and another million to Swansea and Milford. The GWR was facing its own struggles to raise finance to complete Brunel's London to Bristol line, so it was in no position to provide much capital. The shaky finances of the Gloucester & South Wales line extended to not paying Brunel for his work in completing surveys and plans, and in August 1843 *The Railway Times* reported a court case in which he sued the now defunct company for £1,800, the balance of his £3,000 fee. The court ruled in his favour, but it is not known if he was ever paid.[23]

Brunel had to wait almost eight years before the project was revived, this time as the South Wales Railway. The prospectus of the new company argued that the railway would, as well as linking England and Wales with the South of Ireland and its 'commercial and agricultural interests', be best adapted 'for accommodating the extensive traffic of the mineral basin of South Wales'. The route of Brunel's new broad gauge line would form the spine of a railway network which would serve the coalfields, industry and towns and cities of the region[24] and was largely based on an earlier report he had written in September 1836. It was proposed that the railway would leave the Cheltenham & Great Western Union Railway six miles south of Gloucester before crossing the Severn, following the coast and skirting the Forest of Dean. Crossing the Wye at Chepstow it continued to Newport and then on to Cardiff where, the prospectus noted, it would 'unite with the Taff Vale Railway' to access the mineral resources of the area. This was somewhat disingenuous since the TVR had been built to standard rather than broad gauge and so interchange would entail another 'break of gauge'.

Continuing inland through Bridgend and Porthcawl, the railway reached Llanelly by crossing the River Tawe at Landore. The prospectus reported that three-quarters of the copper mined in Britain was smelted at Cwmavon, Neath and Swansea, but initially at least the railway did not run into Swansea. It was planned that the railway would split into two branches beyond Carmarthen, one running south to Milford Haven and Pembroke, and the other north to Fishguard, where passengers and goods could reach Ireland by a six-hour steamer voyage. The completion of the railway would 'create for South Wales and Ireland a new era, and a new character' the prospectus proclaimed, and the 'disadvantages of their present position' would be remedied by the increased traffic of the line ensuring that the 'prospects of renumeration to the promoters of this undertaking are unusually satisfactory'.[25]

There had been considerable opposition inside and outside Parliament to Brunel's plans to shorten the route at the eastern end by building a bridge across the River Severn, instead of starting the new line at Gloucester, which would have been a less expensive and easier option.[26] Brunel proposed a short branch which ran to a swing bridge over the Berkeley Canal, after which the railway crossed the River Severn on a large 1,100 ft bridge at Hock Crib. When his plans became public the business community in Gloucester objected strongly. At a meeting of the Town Council it was resolved to ask the Admiralty and the Board of Trade to investigate Brunel's proposals, which would 'obstruct the free navigation of the river'. The construction of these bridges could not fail to injure the city and port of Gloucester it was argued.[27] In March 1845 James Walker was asked by the Lords of the Admiralty to report on the effects on navigation on the Severn by Brunel's bridges. While observing that their construction would shorten the rail journey from Bristol to Newport by sixty miles, Walker nevertheless concluded that they would seriously interfere with navigation on the river. He also examined an additional proposal by Brunel to build a canal at Arlingham to create a new channel for the river but rejected these as complex and likely to provide extra problems with silting.[28]

There had also been calls in Monmouthshire and from the Duke of Beaufort for the route of the SWR to run further north between Newport and Gloucester via a route that used the Usk valley to reach Monmouth via Caerleon, but this potentially more expensive and steeply graded line was rejected in favour of a more direct route with a branch to Monmouth, which the 1845 Act noted 'would be of great public advantage'.[29]

As a result of these objections, the South Wales Bill finally passed by Parliament in August 1845 allowed only for the construction of a railway from Fishguard and Pembroke Dock to Chepstow; there had been calls during the committee stages to reject the bill entirely. MPs agreed that the issue of the crossing of the River Severn was a matter of concern between the promoters and the City of Gloucester, but the construction of the railway was a question of national interest.[30] The bill also recorded that the South Wales Railway would have a capital of £2,800,000, this total being supported by £600,000 from the GWR who would nominate six Directors. The South Wales was the longest line yet authorised by Parliament at 211 miles and contained a number of potentially challenging and expensive engineering features including a large bridge across the Wye at Chepstow, numerous tunnels and viaducts and harbour facilities at the west end of the line. The Act also specifically mentioned the construction of a 'Drawbridge' over the River Towy at Carmarthen,[31] to be 'constructed in such a manner as the Office of the Lord High Admiral may...approve of', and a similar swing bridge for the passing of vessels was to be provided across the River Loughor west of Swansea.

Given that Brunel's Great Western Railway had finally cost more than £6,000,000 and was only half the length of the SWR, his estimate was on the face of it optimistic. The company had been empowered in the Act to lease or sell the line to the GWR with the permission of three-fifths of the shareholders and so it is perhaps not surprising that as early as August 1845 the SWR Chairman Charles Russell, also Chairman of the GWR, told proprietors that Directors had considered whether it was expedient to pursue some arrangement with the Great Western Railway Company 'for the purposes of drawing still closer the relations between the two companies'. The extensive works proposed by their engineer Mr Brunel 'cannot be accomplished without demands on our resources'. An arrangement would 'give a more steady and certain value to the shares of the South Wales Railway'.[32] Further negotiations took place, and in November GWR shareholders debated a lease of the South Wales by the Great Western with a rent of five per cent on a capital of £3,000,000 in shares, and another £1,000,000 in loans.[33] Any profits generated would be shared after the railway fully opened from between Gloucester and Fishguard. GWR Secretary Charles Saunders told shareholders that the agreement would be fair and equitable and would also induce a 'large class of capitalists' in South Wales to invest in guaranteed stock since previously they might have 'been ignorant of the great capabilities of the line'. A number of questions were asked, including whether atmospheric power would be used on the new railway,[34] but the agreement was unanimously endorsed. There was, not surprisingly, more opposition to the arrangement amongst SWR shareholders, but the difficult financial situation that had affected other Brunel railways, exacerbated by the Irish potato famine meant that there was little room for manoeuvre and the lease was finally approved.

When Russell addressed SWR shareholders in August 1846 he was also finally able to give news of how the company would provide a permanent link with the GWR in the light of the refusal of Brunel's original plan for bridges over the Severn. Undaunted by the verdict of the Admiralty the previous year he had prepared alternative solutions involving a larger bridge or a tunnel, but these, too, had been rejected and reluctantly Brunel finally conceded that a detour through Gloucester was now inevitable. A further Act of Parliament given Royal Assent in July 1846 extended the SWR main line from Chepstow to Hagloe Farm near Awre. Here a new railway ran to Grange Court where it joined the Gloucester & Dean Forest Railway, a local concern which had been acquired by the GWR in 1844 for which Brunel was also named as engineer.[35] From there the SWR had a direct connection to Gloucester and also a connection to the Monmouth & Hereford Railway, another GWR-backed broad gauge line engineered by Brunel. All this was made explicit in the Act, which noted that 'public and local advantage and conscience' would be benefitted if the South Wales Railway 'connected with the Monmouth and Hereford Railway, and with certain collieries and works in the Forest of Dean'.[36] The same 1846 bill also made provision for a branch line to connect with Swansea and Haverfordwest, the former town unaccountably left out of the original bill.

It was reported in August 1846 that 'immediately preceding Mr Brunel's departure for the continent, he entered into very extensive contracts' for the SWR.[37] These took in the construction of the line from Chepstow to Kidwelly and involved work on more than fifty miles of line. To applause from shareholders, the Chairman reported that the contracts were 'in the estimation of Mr Brunel, perfectly satisfactory and advantageous, and they come within the limit of the estimate originally prepared'.[38] Work proceeded, although bad weather slowed progress during the winter months, a seemingly regular occurrence during the construction of many of Brunel's railways. By February 1847 there had been good progress along the line from Chepstow to within fifteen miles of Fishguard.

Brunel reported that the work had been divided into twenty-one different contracts, the heaviest work being done around Swansea, Neath, Newport and Chepstow. It was hoped that if land could be acquired as planned that 'the whole of the works ought to be completed by the summer of 1848' and that if all went well a considerable portion, if not the whole of the line, 'may be opened to public traffic'.[39] Brunel calculated that the cost of the railway was now about £8,000 per mile; with the lease agreed with the GWR, the capital for the company had now grown to £4,000,000, and at a meeting of proprietors held at Paddington in April 1847 he was asked by a shareholder if this total would be enough to construct the line. He replied that he 'had not the slightest doubt that the amount would be fully adequate for the purpose'.[40]

In February 1848 the Chairman reported to shareholders that while good progress had been made on the main line as far as Swansea and Llanelly, work beyond there had been 'retarded' due to shortage of funds. He added that Directors were aware of the advantages of completing the line but they could only make such progress as the finances of the company could justify. Brunel's engineering report noted that the tunnel at Newport was nearly complete and that many of the rock cuttings in the neighbourhood of Chepstow had been excavated, meaning that the line between Chepstow and Cardiff could be opened within the year. Some months later Brunel had to admit to shareholders that the necessity of limiting expenditure had continued since his last report, and he regretted that work on the line had not 'proceeded more rapidly'. Contractors had therefore concentrated on completing works such as tunnels and bridges which 'once commenced, cannot be much delayed'.[41]

Brunel once again made significant use of timber for bridges on the South Wales Railway with nearly fifty of all types being constructed on the line between 1848 and 1856.[42] Two of the largest were at Newport and Landore and in August 1848 Brunel had reported that 'the destruction by fire of a timber bridge across the Usk at Newport will not cause any delay to the works.' Despite clearly wishing to reassure nervous shareholders, the engineer would have been very angry with his contractors. When the bridge caught fire on 31 May it was very close to completion after two years of construction. The Usk Bridge had been built with eleven fifty-feet spans completed with a central span of 100 feet but was destroyed when careless workmen set it alight with a heated iron bolt. The 'Kyanised' timber used in construction was extremely flammable and within minutes the bridge was well ablaze, despite the efforts of both workmen and local people to put out the fire. Although the £20,000 cost would be covered by the contractor's insurance it was a blow to progress. Rather than simply rebuild to the original specification Brunel instead redesigned the central span with an iron bowstring girder structure similar to the one used to cross the Thames on his Windsor branch line. While the Newport bridge was impressive at 650 feet long, the Landore Viaduct was much longer. Crossing the River Tawe which had extensive mud flats on either side as well as another railway, a canal and two roads, at 1,760 feet the viaduct was one of Brunel's most impressive timber structures on any of his railways. The timber bridge had thirty-seven spans and was built by the contractor George Hennet who had already worked extensively for Brunel on the GWR, Bristol & Exeter and South Devon lines.

By 1850 work was almost complete on the first seventy-five-mile section between Chepstow and Swansea. The *Monmouthshire Beacon* reported that the opening had been fixed for 18 June and that 'Mr. Brunel, chief engineer, with several of the Directors, have recently inspected the line.' Every effort would be used, the paper enthusiastically noted, 'to celebrate with brilliancy and spirit an event so auspicious, and so fraught with interest to the inhabitants of South Wales'. The inhabitants of Chepstow had invited the 'engineers and contractors, &c., to dine with them in anticipation of the opening...the Swansea people intend to do "the thing extensive" on the occasion'. This would be a 'grand breakfast' not 'of the ordinary tea, toast, and butter sort', but with the addition of real champagne, the *Swansea Herald* proclaimed.[43] Before the big day, however, there were also reports of a riot as a 'tremendous row took place' between English and Irish navvies near Bishton. Serious fighting took place over two days in a 'spirit of enmity and mischief'. Three ringleaders were arrested and tried, but escaped serious punishment from local magistrates – perhaps because they were shortly to leave and the local population were keen to see their departure without more trouble.[44]

The railway opened as planned on 18 June 1850 and the passage of a special train from Chepstow to Swansea was marked by celebrations on a scale not seen in South Wales for many years. *The Hereford Journal* carried a report a few weeks later that the train was 'besieged with people' at every place. In a speech at Chepstow, a local Magistrate told SWR Directors that the inhabitants of his 'little town' were 'fully convinced, that if we had been left out by you, our situation would have been very different'. At Newport the train was greeted by thousands of spectators and after more speeches extra carriages were added to accommodate more dignitaries. Similar scenes were enacted at Cardiff, Bridgend and Neath, with flags, cannons, fireworks and yet more speeches. At Swansea it was estimated that more than 20,000 people had turned out to see the first train arrive, with Gooch on the footplate of one of two locomotives needed to haul the train. The terminus, although not quite complete, was 'very tastefully prepared for the occasion, the interior and exterior being interspersed with a profusion of banners'.

The Directors and guests must have already had more than their fill of speeches but endured more at a public breakfast for over 400 people. There were numerous toasts from local dignitaries. Mr Talbot MP, the new Chairman of the company, provided a vivid description of early rail travel telling them he had 'been whisked over the tops of trees, and had been so magically propelled through hills and valleys, over and under houses, bridges, rivers, and canals, that the whole of these scenes at Chepstow, Newport, Cardiff, down to Swansea, were so strangely jumbled together in his mind'. It was a great day for Wales he told guests, and 'to the company and to the public he would say on behalf of those who had invested their money in the undertaking "travel, travel, travel"'.[45] Mr Thomas Edward Thomas proposed a toast to the health of Brunel, 'alluding to that gentleman's wonderful achievements in connexion with railway locomotion'. The newspaper reported that the toast was received 'with great enthusiasm, and several rounds of applause' but Brunel responded only with a few words acknowledging the compliment and sat down amid 'protracted cheering'.

Brunel, not always comfortable in such public settings, would have had other things on his mind. For one thing, the SWR was still isolated from the rest of the Great Western Railway at its eastern end through what one contemporary account called 'an unfortunate break' at Chepstow, where Brunel was to design one of his most extraordinary bridges across the River Wye.[46] With the completion of the line from Grange Court in 1851 passengers from Gloucester arrived at a temporary station east of Chepstow and to get to the other side took a bumpy coach ride across the road bridge to the South Wales Railway station. The crossing of the Wye was not a straightforward proposition since approaching from the Gloucester side the railway ran through a steep cutting ending in a sheer drop of one hundred feet to the river. The bank on the other side was a muddy flood plain, so Brunel carried the railway across it on three spans, each supported by a pair of cast iron cylinders that had been sunk into the mud, and gravel to a depth of about fifty feet. These were sunk by heavy weights being placed on them while workmen excavated inside, a difficult and dangerous process as water and fine river silt needed to be constantly pumped out as they dug. Once the work was complete the tubes were filled with gravel and Portland cement, forming a strong foundation for the bridge.

At the west end, the main structure of the bridge was supported by six more cylinders arranged in two rows of three, surmounted with a fifty-foot-high iron tower. On the cliff side, the bridge tower was built from masonry; the bridge itself was an ingenious and striking design that not only took into consideration the unusual topography but also maintained a clearance of fifty feet, enabling ships to pass underneath safely at high tide. Between the two towers were two 312-ft-long wrought iron tubes from which two separate bridge spans were suspended by suspension chains and braced with vertical struts. The tubes weighed only 138 tons each and were manufactured on the side of the river. Early on the morning of 8 April 1852, in an operation supervised by Brunel and his assistant R. P. Brereton and the ever-present Captain Claxton R.N., the first tube was hauled onto a pontoon on the river. The tube was then raised into position, a process that was complete by the following evening. The event had been watched by hundreds of local people including those on two large barges, 'a great number of ladies and gentlemen, amongst whom were Mrs. Brunel and other relatives and friends of the engineer'.[47]

Once all was secured, the railway was able to run trains on the 'down' line from 14 July 1852. The process was repeated after the construction of the second tube in August but the structure was not fully finished until April 1853.[48] The opening of the line from Gloucester to Swansea illustrated the railway revolution perfectly; before Brunel's bridge was built the journey from Swansea to the capital would have taken more than fifteen hours involving a combination of coach, railway and a perilous ferry passage across the Severn, but after 1852 the train trip took just five hours.

In the west, completing the line was a more complex matter, and space precludes a more detailed description of the drawn-out process by which Brunel's line finally reached Pembrokeshire.[49] As well as tapping into the mineral riches of South Wales the promoters of the SWR had also been anxious to develop trade with Southern Ireland. When the lease between the South Wales and Great Western companies was agreed in 1846 it had been intimated that both would help finance two railways in Ireland, the Cork & Waterford and the Waterford, Wexford, Wicklow & Dublin, both to be engineered by Brunel with a view to establishing a through route. However, the desperate situation caused by the potato famine in Ireland meant that by 1847 there was unsurprisingly little enthusiasm for investment in these projects in the wake of such an economic and human catastrophe.

Although Brunel had initially determined that Fishguard was the best place to build a new port for the Irish sea route, the location was far from ideal and heavy civil engineering work would be needed to create a new breakwater and blast rock from high cliffs to build the dock. SWR Directors sought the advice of Brunel's friend and maritime advisor Captain Claxton R.N. and surveys of the coast were made. As an alternative he suggested Abermawr, five miles from Fishguard. This idea was not universally welcomed, although provisions were made in a subsequent Act of Parliament to proceed if required. The wider financial difficulties being faced by the SWR then put paid to the proposal and work was suspended beyond Clarbeston Road, the intended junction for the branch line to Haverfordwest.

The support of the GWR was conditional on the South Wales Railway being completed from Chepstow to Fishguard, and with this an unlikely prospect, the financial position of the SWR was perilous. In August 1849 shareholders were told that the completion of the Fishguard section appeared 'manifestly impolitic' for both companies and Charles Russell who had been Chairman of both the GWR and SWR resigned from the SWR to avoid a conflict of interest. Russell was replaced by C. R. Talbot, the wealthy Member of Parliament for Glamorganshire who took a more active role in promoting the SWR. He also was aware that a new arrangement with the Great Western was crucial while the company pursued its ultimate goal of having a terminus to serve the Irish traffic. By 1851 Talbot was able to tell shareholders that a new agreement with the GWR might be possible, particularly if they could see 'the propriety of substituting the harbour of Milford for that of Fishguard'.[50] This would involve extending the proposed branch for Haverfordwest to Milford Haven instead of striking north west to Fishguard.

With the completion of the Wye Bridge and the opening of the line from Gloucester, a new 999-year lease of the railway by the GWR was agreed, on the understanding that the western terminus of the line would be at Milford Haven. The sheltered waters at Milford Haven were now more attractive to Brunel as he was in the process of planning a third and larger steamship, later to be named the SS *Great Eastern*, which would require a deeper anchorage than previously required.[51] Claxton was again called in by Brunel to help survey the best location for a port in Milford Haven and chose Neyland opposite the naval dockyard at Pembroke. The Haverfordwest branch line became the SWR main line and in 1852 work restarted with the result that by December the following year, the railway from Carmarthen to Haverfordwest was complete as a single-track line. To save money, sections of line used Barlow rail rather than Brunel's baulk road. Barlow rail had a large inverted 'u' shape and was laid directly in the ballast of the track without any fixings. When the track spread under the weight of trains, doubts began to be expressed about its safety and suitability and it was eventually replaced.

With funds still short, the final section of line to Neyland was also only laid as single and the facilities provided at the port were modest. By February 1856 Brunel reported to shareholders that the opening of the line only required the completion of the station, which was then in hand, and three wharves for coastal vessels. The most impressive feature was a floating iron pontoon 150 feet long and forty feet wide, linked to the shore by a bridge. This was brought into use a year after the opening of the line and extended two years later by the addition of pontoons previously used to float out the trusses of Brunel's Royal Albert Bridge. By the standards of earlier openings, the celebrations held to mark the opening of Neyland were muted and Brunel and many other Directors and dignitaries were absent. The port would never be the kind of grand terminus envisaged by SWR promoters in 1844; with the opening of a modern harbour at Fishguard in 1906, it became even more of a backwater but survived until 1964.[52]

In November 1846 Charles Russell, speaking as Chairman of the GWR, told a shareholder that the Vale of Neath Railway, which had received its Act of Parliament earlier that year would be a 'tributary to the South Wales'.[53] This perhaps demeaned the development of the Vale of Neath, an important broad gauge connection for the SWR initially promoted by Henry Coke, Neath's town clerk, a man with a vision to build a new railway to serve the mines and ironworks of the Vale of Neath. Coke travelled to London in May 1845 to a meeting of 'noblemen and gentlemen interested in the formation of a railway from Neath to Merthyr Tydfil'. The meeting resolved that the railway should be constructed to allow 'a more convenient, speedy and proper transit for the extensive mines and minerals of the Vale to the improving port of Neath than at present exists'.[54] It also agreed that the promoters would not follow through with any scheme unless it had the support of the South Wales Railway, adding that 'Mr Brunel be requested to be the engineer to the projected line'. By September a provisional committee had been set up and Brunel had been paid £1,500 for a survey, although much of the actual work was done by Alfred Russel Wallace, later to achieve fame as a naturalist and explorer.[55] Wallace, a capable surveyor, was paid two guineas a day to map out a route through some challenging country. Brunel had already walked the ground some years earlier and presumably had a good idea of where the line might go.

The railway was promoted during the most feverish period of the 'Railway Mania' and much time was spent by Brunel and the Directors dealing with threats from rival schemes such as the East & West Junction Railway and Welsh Midland Railway, neither of which came to fruition. The line was to run from Neath to Glyn Neath before climbing into the mountains through Pencaedrain tunnel at the summit of the line and then running into the Cynon Valley and passing Abernant, turning north east through another long tunnel to reach Merthyr Tydfil. Branches would also run to Aberdare, Penderyn and nearby collieries. Brunel had estimated the cost of the railway at £550,000, but it was the issue of gradients that Brunel was closely questioned about at the committee stage in Parliament. The long and steep climbs proposed were a far cry from the gently graded GWR and he once again found himself facing criticism from his old friend Robert Stephenson, called to give evidence on behalf of the Taff Vale Railway, who were not surprisingly opposed to an incursion by a rival into their territory. Gradients of more than 1 in 50 meant that the difficulties were insurmountable Stephenson said. *Herepath's Railway Journal* reported that the Vale of Neath was 'one of the most extraordinary productions for a railway which ever passed through a committee of the House of Commons', concluding that 'had the line been laid out by anyone but Brunel it would have done for him.'[56]

The engineer was his usual persuasive self and charmed the committee with knowledge of the local area gained from previous surveys and his argument that using the broad gauge would enable more powerful locomotives to be used to haul coal trains. Despite the opposition, and the fact that the railway would be built to his broad gauge, the Vale of Neath Act was nevertheless endorsed by Parliament on 3 August 1846. Having won this battle, the company then found it hard to raise the necessary money and it was reported a month later that 5,000 shares were as yet unallocated. Brunel was instructed to make a start and contracts were let in 1847 on the easily graded nine-mile section just out of Neath. With money short, work did not start on the Merthyr tunnel until February 1848. That year the slow pace at which work was then progressing required permission from the Railway Commissioners to extend the time required to complete it. When two contractors excavating the Merthyr tunnel failed, the board decided to concentrate on completing the line to Aberdare. It had been intended that the station there should be built close to the Taff Vale Railway station with interchange facilities, but relations between the two companies were so poor that Brunel chose a more distant location on the other side of the river for the Vale of Neath station.

The railway was provided with a single line of broad gauge track rather than the planned double track formation as costs increased and the completion date receded. The contractor Ritson, having excavated the Pencaedrain tunnel then moved on to the 2,495-yard Merthyr tunnel in February 1851, but work was not finished there for another two years. In September 1851 only 680 yards of the tunnel had been dug. There was some good news for the company that month when the line finally opened to Aberdare, although the opening had been delayed by wet weather which had caused landslips. The special train hauling a party of Directors ran along the line to Aberdare on 23 September. It struggled on the climb from Glyn Neath, but a good lunch was enjoyed by the group when they reached their destination, even though, as one observer noted, most of the stations were 'but partially erected'.[57]

There was criticism of both Brunel and the slow pace of construction from shareholders and Directors. In 1850, H. A. Bruce, a member of the Vale of Neath board, announced that he would recommend at the next meeting that Brunel's salary be reduced from £1,000 to £600 per year in view of the 'unreasonable time' it was taking to complete the railway; absences as work on other projects took him elsewhere did not improve the opinions of his opponents, although his appointment diaries for 1850 and 1851 show many hours spent in meetings about the railway both in London and Wales. An entry for Wednesday 15 January 1851 recorded that Brunel 'spent the whole day on Vale of Neath railway business'. Having arrived by mail train from Gloucester he travelled by post chaise to Chepstow, where he caught a SWR train to Neath, arriving there at 7.30am. Brunel 'met with Mr R. W. Jones and went along the Vale of Neath railway line with him to see the slip and the general works'. Having visited Aberdare, he then caught a Taff Vale train from Merthyr to Cardiff, ending the day back at Chepstow where he dined with Lord Villiers, the Chairman of the Vale of Neath.[58]

Meanwhile, navvies in the Merthyr tunnel struggled on and while excavations were reported as being almost finished in February 1853, the line to Merthyr was not finally opened until November. Timber viaducts and bridges were once again used by Brunel on the Vale of Neath, both on the main line and the Dare and Aman branches opened in 1854 and 1857 respectively, the viaducts there only being replaced in 1947. Branch lines were also laid with Barlow rail, but it was as unsuccessful there as it had been on the South Wales Railway and was eventually replaced. Coal trains began running from Aberdare in 1852 and for these Brunel designed special flat trucks which could carry four iron containers filled with coal. These could then be lowered into the holds of colliers at the docks where hinged trapdoors could be opened allowing the coal to fall more gently with less breakage, since the coal mined in the Vale of Neath was more friable than elsewhere in the South Wales coalfield. The success of this idea does of course beg the question why Brunel could not have made more effort to design similar goods wagons for use elsewhere on his broad gauge network, making the 'break of gauge' a less onerous problem.

The completion of the Vale of Neath gave the South Wales Railway access to the valuable mineral resources it had claimed would make it a profitable line when it was first promoted in the 1840s. In 1854, however, the Chairman reported that 'coal traffic along the line had become extremely important [but] regretted to say that it was by no means as extensive as it might be.'[59] The reality was that being built to the broad gauge, Brunel's SWR main line was isolated from the standard gauge railways like the Taff Vale which had been built to run coal from pit to port and most of the hoped-for coal traffic went direct to docks like Newport, Cardiff and Barry. It would not be until the grouping in 1923 that the GWR, which had taken over the South Wales Railway sixty years earlier, would reap the benefits of South Wales coal trade.

9

The Final Years

By 1850 the frantic pace of the Railway Mania had subsided and the wild speculation of investors was replaced by a recession and shortage of capital felt keenly by many of the railways authorised by Parliament in 1845 and 1846, and the GWR itself. This process, which had begun some years earlier, had been exacerbated by the fall of George Hudson in 1849 which had left many shareholders and financiers in trouble as the value of railway stocks tumbled.[1] Brunel had also been forced to dismiss some of his own staff due to 'the state of trade and money in England' at that time. Writing to his brother-in-law John Horsley, he confessed that his spirits had been 'positively broken' by the process of cutting salaries and sacking staff. 'Imagine the situation of unfortunates – and the extent of disappointment pain and misery I have had to inflict.' His comments were in contrast to some of the sharper treatment he meted out to staff in earlier years. The letter certainly demonstrates disenchantment with his situation and perhaps a world-weariness induced by more than fifteen years of almost constant toil: 'All my life is one of slaving and compulsion,' he told Horsley.[2]

The 'anxieties and vexations' Brunel reported in his letter had been made worse by the embarrassing failure of the atmospheric system on the South Devon Railway and also the stranding of his beloved SS *Great Britain* in Dundrum Bay in September 1846,[3] as well as the debate and rancour of the Battle of the Gauges that was still rumbling on. Brunel's fragile health, no doubt tested by the gruelling work regime he had endured since 1833, may have been uppermost in his mind; as he began his fourth decade there is some indication that while he was not considering retirement, he was thinking about a time when he might work less. Despite the economic challenges of the 'Hungry Forties' felt by many in England and the difficulties facing railway companies more generally, Brunel's own financial position nevertheless appeared secure enough for him to purchase the property next to his London residence at 18 Duke Street in 1848, the combined complex giving the engineer and his family a grand home and providing him with expansive office space.

Extending his luxurious house in London was not his only ambition; having spent many months in Devon surveying and supervising work on the ill-fated South Devon Railway, Brunel was also anxious to acquire a country estate in the area where he could retreat, away from the hustle and bustle of the capital. Brunel and his family had previously stayed in a number of locations in Torquay, but in 1848 he purchased land at Watcombe where he planned to build a house. For the next few years Brunel and his family spent most summers in Devon, living in a rented houses for the season since Torquay was also a suitable location for the engineer to both rest and travel to the various railway projects he was still engaged on in the West Country, the Midlands and Wales. The office diaries show frequent absences in 'Devonshire' from 1850 onwards, in striking contrast to earlier years. Despite the continuing pace of his working life, the planning of the estate at Watcombe provided a welcome relief from his engineering work and surviving notebooks and correspondence about planting and the design of his proposed new home demonstrate the same level of detail and commitment he showed in his railway and maritime projects.[4]

The end of Railway Mania and the ensuing lack of investment capital certainly put an end to the frenetic pace of railway development in which Brunel had been an integral part. Much still remained to be done, however, and in the first half of the 1850s in particular, he was still busy on railway projects, most having been authorised in the years just prior to the Gauge Commission. With no more broad gauge lines in prospect, Brunel was engaged in a mopping-up operation to complete railways which still competed for territory with standard gauge rivals north, south and west of his original London to Bristol line.[5] In addition, as work began to dry up, he looked abroad for commissions[6] as a consultant, although as a number of historians have pointed out, he would not have approved of this term. His office diaries also suggest a change in the way in which he spent his working week, Brunel spending more time at Duke Street with long days of meetings which often began early in the morning and ended after midnight. This gruelling regime was broken with longer trips away from the office over several days when he visited the 'distant works' of his railways to check on progress. Isambard Junior recorded that 'It was impossible for Mr Brunel to look after all his works to the same extent as he had done in the case of the Great Western Railway,' adding that it was a 'great evil' that engineers were 'prevented from attending properly the construction of their works'[7] by appearances in parliamentary committees. These took an increasing amount of Brunel's time, both in support of his own engineering projects but also giving evidence in other railway enquiries.[8]

A significant piece of unfinished business was the construction of a new station at Paddington. The difficulties faced by Brunel and the GWR in the 1830s in finding a central location for its London terminus meant that by the time they had finally settled on Paddington there was not the money to build a grand station on the scale of the London & Birmingham Railway's Euston. Since it was not Brunel's first choice of location, the first station at Paddington was built in a hurry and lacked the refinement of the kind of elegant terminal he had in mind. It opened in 1838 a few months after the first trains had run between London and Maidenhead and consisted of two arrival and two departure platforms situated on land between two bridges, one carrying Westbourne Terrace at the west end, and beyond the buffer stops to the east, Bishop's Road. The arches under Bishop's Road bridge formed the entrance to the station, and contained waiting rooms, a booking hall and other offices. Passengers were protected from the elements by a modest wooden overall roof supported by slender cast iron columns. There was also a small goods shed and at the west side of the station a locomotive shed and workshops, where Daniel Gooch spent many hours in the early days of the railway working to keep the unreliable locomotive fleet running.[9]

When the station was opened in 1838 the train service was limited to just nine trains each way, but as the line was extended further west, this number increased; by the time the railway was opened through to Bristol there were seventeen trains running in each direction. From March 1840 Third Class passengers were 'conveyed in uncovered trucks by the Goods Trains only'[10] the company announced, and two additional trains were run daily for this purpose. Facilities became more and more unsatisfactory as new lines were added to the network bringing more passengers; the opening of the Bristol & Exeter, Cheltenham & Great Western Union railways and new branch lines all put additional pressure on the small terminus. With capital still in short supply, the GWR board remained reluctant to commit to the cost of building a larger terminus, but with the prospect of other railways such as the South Wales and Oxford, Worcester & Wolverhampton line opening after 1850 GWR Directors finally conceded that a new station was needed. Brunel began to sketch out ideas towards the end of 1850.

Matters then moved quickly. At the Half-Yearly Shareholders meeting in February 1851 it was reported that the board had agreed that 'the time has arrived when a commencement must be made in the construction of more permanent buildings, fitted for the great increase in business.'[11] Temporary additions to the existing station had been 'maturely considered' but this, the board argued, would not be 'consistent with economy' and as a result the construction of a new station was agreed at a cost of £50,000. In 1836 Brunel had sketched out plans for a station in a form similar but larger than his terminus at Bristol, with passengers entering a concourse opening into arrival and departure platforms separated by a roadway with two discrete train sheds. When he revisited the plans, his initial proposals retained separate arrival and departure platforms, but they were completely covered by a roof. The size and scale of the new station precluded the use of timber which had been so successful at Temple Meads and Brunel turned instead to the use of iron and glass for his new roof.

Brunel was not the first railway engineer to design a large overall roof for a station[12] but his work at Paddington owed much to the development of the 'glasshouse' technology which had been demonstrated at the Crystal Palace at Hyde Park, which housed the Great Exhibition of 1851. The roof there had been the work of Joseph Paxton, Head Gardener at Chatsworth House in Derbyshire whose invention of a 'patent glazing system' revolutionised the construction of large iron and glass buildings.[13] Brunel had been an important member of the building committee for the Great Exhibition along with William Cubitt and Robert Stephenson when it was established in February 1850. With an opening date of May 1851 already set, the committee considered entries in a competition to design the building but with no suitable winner, they produced their own plans for a large structure that looked very like a railway station, complete with a massive dome designed by Brunel. The ugly design was not popular and the whole project was rescued only by the last-minute contribution of Paxton. He produced a design supported by a tender from the contractor Fox, Henderson & Co, to construct the exhibition building. The design was accepted in July 1850. Brunel remained closely involved with the construction of the Crystal Palace, his appointment diaries recording many meetings relating to the project in 1850 and 1851.

Although some complained that the main structure of the Crystal Palace was monotonous in form and lacking in detail and colour[14] there was much Brunel could take from the design. The building had been erected in seventeen weeks and was a masterpiece in pre-fabrication with components manufactured elsewhere. Brunel had seen at first hand the way in which cast-iron columns and beams were assembled and placed in position and the methods used to fix the 900,000 square feet of glazing into the roof by the contractor Fox, Henderson. The Crystal Palace was known as much for the way it was built as for its final form.[15] The site of the new station was to be constructed east of the Bishop's Road Bridge, then occupied by the old goods depot in a space bounded by Eastbourne Terrace and Conduit Street[16] and since it was situated in what was essentially a cutting, it would be 'all interior and all roofed in', with little outside ornamentation.

Brunel's original sketch plan showed a station entrance from Conduit Street leading to a concourse at the head of the platforms. This idea was changed following concerns expressed by Directors about the lack of suitable hotel accommodation at Paddington. A private establishment, the Prince of Wales Hotel, had been located close to the old station at Bishop's Bridge but it was far from satisfactory, one writer describing it as 'nothing more than a Public House with rooms'.[17] Prompted by the prospect of thousands of people descending on the capital for the Great Exhibition the following year, one of the GWR Directors James St George Burke wrote to Charles Saunders in September 1850 arguing that 'the existence of a good hotel at the end of a journey is such an inducement to many travellers.'[18] At a board meeting in December 1850 Brunel produced plans for Paddington that would enable the Company to determine 'the most eligible position for building the proposed hotel'.[19] With the need for a hotel uppermost, the board rejected Brunel's idea for a grand entrance and decided that his station would instead sit behind a hotel situated on Conduit Street. In addition to this rebuff, the company also made the bold decision to appoint the architect Philip Charles Hardwick to design the hotel. The speed at which this decision had been taken can be judged by the fact that Hardwick then appeared at a board meeting on 2 January 1851 and by 9 January had produced plans and estimates for approval. On the face of it, one might have assumed that Brunel would not have been pleased to have been passed over for this work, but there is no surviving correspondence to suggest he complained about the situation. The reality might well have been that Brunel was simply too busy to design the hotel given that he was still completing railway projects in Wales, the West Country and the Midlands and a myriad of other schemes including the newly formed Eastern Steam Navigation Company, set up to build the SS *Great Eastern*.

This heavy workload appears to have forced Brunel to adopt a different and – unusually for him – more collaborative approach to the building of Paddington using contacts made during the building of the Crystal Palace. His involvement in the exhibition had brought him into contact with Matthew Digby Wyatt, an architect who was the secretary of its executive committee. Digby Wyatt had designed a 'Moorish pavilion' which Brunel had admired. On 13 January 1851 Brunel wrote to Digby Wyatt telling him 'I am going to design, in a great hurry, and I believe to build, a Station after my own fancy.' The station would be roofed in and made entirely of metal he added, and it 'almost of necessity becomes an Engineering Work, but, to be honest if it were not, it is a branch of architecture of which I am fond, and of course, believe myself fully competent for'. He admitted that he did not have time or knowledge for 'detail of ornamentation' and asked if Digby Wyatt would be willing to work for him in the 'subordinate capacity' as his assistant. Brunel reminded Digby Wyatt that working for him would be 'as good an opportunity as you are as you are likely to have' and asked him to attend a meeting at 9.30pm the very same evening. Digby Wyatt was otherwise engaged but visited Brunel the following day.

Impressed by their work in completing the Crystal Palace under pressure, and aware of their extensive experience in working with iron, Brunel was anxious to engage Fox, Henderson & Company as the contractor to build Paddington. In such exceptional circumstances he was able to persuade GWR Directors not to seek additional tenders but to accept them as sole contractor. This was duly agreed, and Brunel then worked with Fox, Henderson, Digby Wyatt and T. E. Bertram, one of his long-standing resident engineers on the original GWR line, to develop the plans. Brunel as usual provided sketches for outline designs and plans but instead of detailed drawings being done by his draughtsmen at Duke Street, this was done by the Fox, Henderson Company.

Brunel's new station consisted of a train shed of three elliptical arches, a central span of 102 feet 6 inches, with two side spans of 69 feet 6 inches and 68 feet respectively. Each span had sixty-three wrought iron ribs supported by cast-iron columns; cathedral-like transepts broke up the flow of the 700-foot-long structure, also bringing more light into the station. Contracts for the construction work were let in two stages in 1851 and 1852, but progress was slow, no doubt due to the fact that Fox, Henderson had been contracted to move the Crystal Place from Hyde Park to Sydenham after the Great Exhibition had closed. Directors were frustrated at the lack of progress and matters were not helped when Brunel became involved in a dispute with them about his expenses in March 1852. Brunel had personally employed Digby Wyatt, a move which he argued had saved the Company money and had also employed draughtsmen and assistants which had cost him over £1,000 of his own money.[20] Directors seemed unaware of this private arrangement and denied any responsibility for paying Digby Wyatt with a stern admonition that they expected this kind of architectural work to be Brunel's responsibility. The Directors were also annoyed when they discovered that Brunel had also engaged the designer Owen Jones to advise him on colour schemes for the roof. They were less than impressed when they visited the station in April and saw a sample area of the roof painted in an elaborate style and Brunel was told in no uncertain terms by Charles Saunders that they wanted the station to look as plain and inexpensive as possible, and the idea was dropped.

By early 1853 the departure side had not been completed and with Brunel busy on other projects the Directors ordered Saunders to write to him to express their displeasure; in April Fox, Henderson received a stronger ultimatum threatening their dismissal. This finally prompted some action and by the end of the year the departure side roof was complete and after delays due to bad weather in December, trains finally left the station on 16 January 1854. The main roof span was completed soon after and trains began arriving at the new Paddington in May. Hardwick's hotel opened the following month, but work continued on associated parts of the station for more than two years on the goods depot and engine and carriage sheds. By 1857, the construction of Brunel's great station had cost the Great Western railway more than £660,000. The combination of Brunel's original conception and plan and Digby Wyatt's ornamentation had, however, finally provided the company with a London terminus it could be proud of, and it remains as one of the engineer's greatest achievements.[21]

Writing in November 1854, Brunel recorded that he had engineered 1,046 miles of railway since the first contracts had begun on the Great Western eighteen years previously adding that with the opening of the 'Birmingham & Dudley' this total would include around 500 miles worked by the GWR and the rest 'under my direction'.[22] Access to Britain's second city Birmingham and the construction of a terminus there was a complicated story linked closely to the Battle of the Gauges and the ambition of Brunel and the GWR to push broad gauge lines further north to Liverpool. The line referred to by Brunel was in fact the Birmingham, Wolverhampton & Dudley Railway (BW&DR) authorised by Parliament on the same day as the Birmingham & Oxford Junction Railway B&OJR) on 3 August 1846. The BW&DR was an independent company with its own engineer John Maclean, built to connect Birmingham with the OW&WR near Bilston, though its Directors included Francis Rufford, Chairman of the Worcester line and a number of GWR Directors such as Frederick Barlow and William Tothill. The Birmingham & Oxford Junction Railway had originally been promoted and encouraged by both the Great

Western Railway and its rival the Grand Junction Railway as direct competition to the London & Birmingham (later absorbed into the London & North Western) and was to connect with the GWR at Fenny Compton on the Oxford & Rugby line. After some wrangling, the GWR took over both the B&OJR and Birmingham, Wolverhampton & Dudley companies in 1848.[23]

Not surprisingly, the standard gauge London & North Western Railway was fiercely opposed to broad gauge trains running into Birmingham and rejected the idea of a joint station. What followed was a final skirmish in the Gauge War when both companies rehearsed familiar arguments once again. Questions were supplied by Board of Trade Railway Commissioners to both companies with regard to the gauge and the operation of trains. Particular attention was paid to the question of the 'mixed gauge', a formation of three rails that allowed the passage of both broad and narrow gauge trains on the same track. In his evidence Robert Stephenson argued that the system was inherently dangerous, particularly where points and crossings were concerned, but the Commissioners did not agree and ruled that the broad gauge could be extended to Birmingham on the basis that it was a mixed gauge route. Gooch recorded many years later that this was 'the last of the real broad gauge fights',[24] although a further ruling from the Commissioners noted that the broad gauge should not be extended beyond Wolverhampton until the lines had been thoroughly tested.[25] The GWR had already conducted experiments with mixed gauge track and were confident that it would work, and so it proved, although in the long term its advent marked the beginning of the decline of Brunel's revolutionary innovation. Although a single broad gauge track was opened from Oxford to Banbury in 1850, the line to Birmingham Snow Hill was not completed until October 1852, when passengers could finally travel by broad gauge from Paddington to Birmingham. The new station in Birmingham was not a terminus, as trains could continue on to Wolverhampton, although this stretch along the route of the original BW&DR was not opened until 1854. Although engineer of the Birmingham & Oxford Junction Railway, ironically it seems that Brunel had little influence on the line that regularly used broad gauge trains, the BW& DR, and much less appears to be known about his work in the West Midlands area generally.[26]

Brunel also found himself involved in the completion of a number of lines south of his Bristol to London main line which had started life during the Battle of the Gauges, but whose completion dragged on well into the 1850s. In the south the GWR had been pitted against another formidable adversary, the London & South Western Railway, who, like the L&NWR, opposed both the broad gauge and any attempt by the Great Western to push into their territory. In 1844 the L&SWR had proposed a new line that would run from Basingstoke into the heart of GWR territory at Swindon; in retaliation, the Great Western announced the construction of the Berks & Hants, a railway from Reading to Basingstoke and Newbury, and the Wilts & Somerset Railway, an ambitious scheme that would run from Corsham near Chippenham south to Salisbury, with branches to Devizes, Frome and Bradford on Avon and what was called a 'coal line' to Radstock. With connections at Salisbury this route would, it was hoped, provide the GWR with a more direct route to Southampton and the south coast.

The final link in this new route which would run deep into L&SWR territory was a proposal by the Bristol & Exeter to build a line from Taunton to Yeovil and then on to Weymouth. This plan stalled in the autumn of 1844 when The B&ER changed its mind, agreeing only to run to Yeovil and so the route of the Wilts & Somerset was changed to take in Yeovil and Weymouth with branches to Sherborne and Bridport. This grandiose scheme was then renamed the Wilts, Somerset & Weymouth Railway. The L&SWR had also been promoting rival schemes in Devon and Cornwall against the interest of the GWR and as a result the competing proposals were considered by the Board of Trade, who after some debate ruled in favour of the GWR and its broad gauge lines. An agreement was struck between the two companies 'not to encourage or promote, directly or indirectly, any future line of railway in opposition to, or tending to divert legitimate traffic'.[27] In order to 'preserve friendship between the companies and to avoid unnecessary contention' it was agreed that the GWR would not extend its lines south of Salisbury or Dorchester and the L&SWR would not promote railways westwards from the same place. They also agreed to withdraw their opposition to broad gauge railway lines proposed in the far west. This agreement was not quite the end of the battle between the two companies, but with a truce in place, bills for the Berks & Hants and Wilts, Somerset & Weymouth and the Bristol & Exeter's branch from Taunton to Yeovil had an easy passage through Parliament.

Of the three lines authorised, for Brunel the Berks & Hants proved to be the most straightforward. From the opening of the GWR main line in 1840 there had been calls to build a railway to Newbury and *The Berkshire Chronicle* had reported in 1842 that there had been many enquiries as to whether there was 'any serious intention or contemplation'[28] of such a line, suggesting that the Mayor call a public meeting to drum up support. When the railway company was proposed it was set up as an independent company, although the list of eight original Directors reveals some familiar names including Frederick Barlow and Thomas Guppy, so it was no surprise that the Berks & Hants was absorbed into the Great Western Railway in May 1846. The B & HR was a short twenty-five mile railway which ran from Reading through the Berkshire countryside to Newbury with a fifteen-mile branch to Basingstoke. Additional powers were sought in 1847 to extend the main section from Newbury to a temporary terminus at Hungerford. Brunel estimated the cost of the line at just £400,000 as there were few engineering challenges to overcome in comparison to railways he had engineered elsewhere. When the Board of Trade Inspector visited the railway in December 1847, he noted that 'there were many wooden bridges under and over the line, some of considerable size.'[29] Brunel making good use of timber as usual.

The railway was opened to Hungerford on 21 December 1847, the station there built to Brunel's 'one-sided' pattern with a temporary wooden building provided in the first instance. A further bill for the Berks & Hants Extension Railway which would run from Hungerford to link with the Wilts, Somerset & Weymouth line was promoted in 1846, but following a protracted struggle in Parliament the proposals were shelved and not revived until August 1859, a few months before Brunel's death. The branch to Basingstoke opened in November 1848, progress slowed by difficulties with the LSWR. The only intermediate stopping place was provided at Mortimer; Brunel had been constrained by an edict from the Duke of Wellington, who insisted that no station be built within five miles of his house, Stratford Saye, without his consent. Following pressure from neighbours he allowed one to be built at

Mortimer three miles away 'as commodious as His Grace might require'. Brunel duly provided a pretty, Italianate brick chalet-style station building that survives today.[30]

The Wilts, Somerset & Weymouth Railway was promoted by the GWR from the beginning and Brunel had attended a public meeting in Warminster in July 1844 to share his plans for the railway, which would run from the GWR main line at Thingley Junction to Trowbridge and via the Wylye Valley to Salisbury. The complex nature of the railway and its subsequent history can be gathered by the fact that the original Act of Parliament passed in 1845 listed no less than seven separate lines to be constructed. A further change was an extension of the railway from Bradford on Avon through the Avon Valley to join the GWR main line at Bathampton, which Brunel argued would be 'advantageous' to the company.

At the first General Meeting of the Company held in October 1845 the Chairman Walter Long MP argued that he had not remembered 'a Bill passing through the House of Commons as easily as this one did'. His speech concluded that the line when completed would be 'ranked amongst the first in the kingdom' and he confidently expected that the line would not only be 'a good one for the public, but a remunerative one for its promoters'. Shareholders were also told that the route of the railway had been re-surveyed and the line staked out from Thingley to Salisbury. Contracts had been agreed for the whole line with 'experienced contractors' and should be let with as little delay as possible.[31]

Brunel estimated that the railway would cost £1,500,000 although the Act provided the WS&WR with the option to borrow a further £500,000. In 1845, the Chairman told shareholders that the company had 'plenty to go on with for the present' and a further call on funds would not be required, but his optimism was misplaced and although following Brunel's survey the railway was able to acquire land and appoint contractors, the financial position of the WS&WR worsened. By September 1848 it had only been possible to complete thirteen miles of line from Thingley Junction to Westbury with 120 miles of railway left unfinished. Contractors were let go, and earthworks, bridges and permanent way lay unfinished for years. Much of the trackbed to Bradford on Avon had been completed by 1848, but rails were not laid until 1856 when the line was connected at Bathampton. By 1850 it was clear that the small company were unable to continue. It was formally absorbed into the Great Western Railway in July 1851. Even the GWR struggled to find enough money to finish the railway as planned, leading to legal action from residents in Bradford and Devizes to enforce completion.[32]

Brunel maintained his position as engineer during the protracted construction of the railway but it is likely that he relied on assistants to undertake detailed design work on the WS&WR including the fine wooden train shed at Frome, which, although built in his style, was the work of J. R. Hannaford.[33] Under the management of the GWR work slowly progressed, with the Frome-Westbury section opening in October 1850 and Warminster to Westbury complete the following year. The final section to Weymouth was not opened until January 1857, by which time Brunel's attention was firmly fixed on work in Cornwall and at Millwall, where his giant ship the SS *Great Eastern* was close to completion.

Despite his workload, and the slow progress of the WS&WR, Brunel still found time to take on another commission in 1855 when proposals were submitted for a railway between Frome, Shepton Mallet and the ancient cathedral city of Wells. The Great Western Railway assured Directors of the fledgling company that they had regarded 'with much interest and favour the promotion of a broad gauge line of railway between Frome and Wells'. This was of course provided there was a direct connection to its network and on the understanding that 'no pecuniary assistance of any form' was requested from them.[34] The East Somerset Railway as it became known, was completed in November 1858.

Whatever feelings the residents of Wells may have had concerning their relative isolation in rural Somerset and their lack of railway communication, it was nothing compared to the population of Cornwall, the last county in England to have a railway link with the capital. The sparse population and challenging landscape of the Royal Duchy had made it a risky commercial proposition for railway companies, but the need for a railway had nevertheless been keenly felt locally since the 1830s. Well before Brunel arrived on the scene, a number of Cornish railways and tramways had been built to carry the products of mines and quarries,[35] but the county still remained a region 'beyond railways'.[36] By the 1840s, with the completion of the Bristol & Exeter Railway and the construction of the South Devon Railway, Brunel's broad gauge route from Paddington to the far west was nearing completion and interest in the construction of the final link, a railway through Cornwall, was revived.

The construction of a railway across the Royal Duchy to Penzance would eventually involve the building of two lines by the Cornwall and West Cornwall companies, both of whom received Royal Assent for their schemes on 3 August 1846 and both of whom would employ Brunel as their engineer. The original prospectus of the Cornwall Railway had been issued in 1844 following a survey not by Brunel, but by Captain William Moorsom, an experienced civil and railway engineer who been involved in Cornish railway schemes since 1840.[37] Moorsom surveyed a route which ran from Plymouth to Torpoint and on to Liskeard, Par, St Austell and Truro and it was this proposal which was submitted to Parliament in 1845.

The GWR and its 'Associated Companies' had offered financial support for the proposal but to the surprise of the Cornwall Railway promoters, the development of a new line became a battle between railway companies as well as a matter of local interest. The L&SWR revived its Central Cornwall railway scheme, in competition to the more southerly Cornwall railway route, which connected with Brunel's other broad gauge lines, and submitted a bill to Parliament themselves. As *The Railway Times* would report later, competition between the two big companies for Cornish business was not merely a consequence of the Railway Mania but was a more commercial fight based on the fact that passenger receipts on the Cornwall line might amount to £500,000 per annum and mineral traffic from mines could generate more than £1,500,000 per annum. This was something that had been discussed by the 'influential men of the county for the last ten years' it concluded.[38]

The L&SWR proposal was rejected by Parliament largely on the grounds of the engineering difficulties that the line would have presented. Although it skirted the edge of Bodmin Moor to access the towns of Liskeard, St Austell and Truro, it still crossed some challenging terrain, contemporary sources noting that there was scarcely a yard of line that would have been built on natural ground, meaning that many cuttings, embankments, tunnels and viaducts would have been needed, making the railway both expensive and difficult to operate.[39] Moorsom's plans did not fare much better in the committee stage either, as his route contained its own fair share of steep gradients and sharp curves, explained away by the promoters' claim that they would use the as yet unproven atmospheric system on large sections of the line. Just as problematic was the suggestion that to cross the River Tamar the railway company would use a train ferry, something which committee members in Parliament were dubious about.

Brunel, as engineer of the GWR and the Associated Companies, gave evidence in the committee stage of the Bill. Members clearly thought this a slightly odd situation, one noting that there had been 'some misapprehension...as to the nature of your connection with this railway'. Confirming that he was not the engineer of the Cornwall line, he told the committee that the survey of the route had not been done under his supervision. All the work had been done by Moorsom, although he had consulted Brunel occasionally on 'one or two little points'. Clearly reluctant to criticise Moorsom directly, Brunel answered questions on the operation of a train ferry, the use of the atmospheric system and the gradients to be encountered on the line, giving sometimes non-committal answers. He told the committee that he had spent time looking over the route of the line but had not been professionally employed to do so, but added that 'the Cornish line is something I have looked forward to for some time.'[40]

The House of Lords rejected the Cornwall Bill with a suggestion that they would take a more sympathetic view in future if the matter of the train ferry and the reduction of the severe gradients proposed by Moorsom were reinvestigated with a 'further and most accurate survey of the whole country traversed between Plymouth and Falmouth'.[41] It comes as no surprise then that Moorsom's services were dispensed with and Brunel was asked to become engineer of the railway. In September 1845, shareholders were told that following a meeting of the Cornwall and West Cornwall Railway committees and representatives of the Associated Companies, Brunel had been asked to 'carefully survey the country, for the purpose of recommending such improvements or alterations in the line of railway as may seem to him expedient'.[42] Brunel's revised survey produced a new proposal that followed much of the route originally proposed by Moorsom but had much easier gradients that did not exceed 1 in 60. The Tamar was to be crossed not by a ferry but by a bridge that would then connect the line with St Germans and Liskeard.

A new bill was submitted to Parliament and received Royal Assent on 3 August 1846. The Cornwall Railway would provide 'great public advantage to connect the towns of Plymouth and Falmouth – facilitating an expeditious means of communication', the preamble stated. The company had a capital of £1,600,000, higher than the original scheme, with the Great Western, Bristol & Exeter and South Devon railways [43] subscribing more than £300,000. The railway was to be broad gauge and the Act also stated that the bridge over the Tamar should consist of four spans only, with dimensions, height and construction approved by 'The Lord High Admiral'. The Act also ordered that rent of £25 per year was to be paid to the estate of 'His Royal Highness Albert Edward, Prince of Wales in right of his honour and manor of Trematon' on account of his entitlement to 'the Water of Tamar'.[44] Having won the battle to build the railway and made tentative steps towards commencing construction, work was then suspended due to the prevailing economic conditions. The company returned to Parliament and a further Act was passed on 20 December 1847 'to give further time for making certain railways' and work ceased for three long years.

In the meantime, the West Cornwall Railway, which had received its Act of Parliament on the same day as the Cornwall Railway, were anxious to proceed despite the lack of progress shown by their neighbour. The line had been originally promoted by the business community in Penzance who wished to end its isolation from the rest of the country and connect it with the rest of the railway network. The West Cornwall Railway was established in 1844 to link Penzance and Truro, with branches to St Ives, Penrhyn and Falmouth. Part of the route would take in part of the Hayle Railway, an existing standard gauge line and on steeper sections of the railway between Redruth and Truro, trains would be worked on rope-hauled inclines.

Once again much of the initial survey was carried out by Moorsom, although Brunel's influence seems much in evidence, especially since in addition to the inclines, atmospheric traction would also be used as well as locomotive power, all in a twenty-five-mile stretch. When the West Cornwall Bill reached the committee stage in Parliament in 1845, after discussions about the commercial prospects of the railway it was Moorsom, not Brunel, who faced questions from peers about the engineering and route of the line. Given events elsewhere in Devon it was no surprise that Moorsom was asked about the suitability of the atmospheric system, but he appeared confident that it would be suitable for the line having seen it in action in Ireland. Questions were also asked about the gauge of the railway, and since part of the route surveyed would incorporate the Hayle Railway, built to Stephenson's 4 feet 8 ½ inch gauge, Moorsom stated that the West Cornwall could also be built to this gauge, adding that there would be no difficulty in converting the line to broad gauge at a cost of £500 per mile.

As was the case with the Cornwall Railway Bill of 1845, the Lords threw out the West Cornwall scheme, the meagre Moorsom estimate of £180,000 and over-complicated combination of atmospheric, rope-hauled and locomotive haulage proving too much for the peers. The luckless Moorsom was once again ousted and Brunel was asked by the West Cornwall to undertake a new survey which removed the troublesome inclines. A new Bill was submitted in 1846 and in the committee stage Brunel was questioned about the new route between Truro and Penzance. After some probing he was forced to admit that he had not actually undertaken the survey since he had been away in Italy; the survey had been done by one of his assistants, William Johnson, and Brunel had in fact only spent a couple of hours in West Cornwall. This, however, he told peers unashamedly, was more than enough time to cover the ground! The bill was nevertheless approved with some amendments[45] and a capital of £500,000.

There had been little progress by the summer of 1849 and Brunel was asked to advise on less expensive options. He recommended building the line as a single track railway and using Barlow Rail to reduce the cost to £292,628; this news was announced to shareholders by Captain Moorsom, who was now Chairman of the company, but a year later there still had not been much progress. Moorsom's tenure appears short-lived, as the WCR board and their new chairman H. O. Wills were forced to take more drastic measure by announcing that the line would be laid to the standard gauge, a move confirmed by Parliament the following year. Significantly, however, there was a proviso in the legislation that the WCR should lay broad gauge track if requested at six months' notice by a 'connecting railway'.

Finance was finally put in place in early 1851 to build the railway from Hayle to Penzance and divert the route of the line around the steep inclines on stretches of the old Hayle Railway. The contractor Ritson was engaged to undertake this work, and subsequently asked to complete the remaining section of line to Truro. The remote location of the railway meant that rail, rolling stock and other materials had to be delivered by sea, causing some delays, but by the spring of 1851 more than 600 men were at work. In March 1852 the railway between Penzance and Redruth was opened and following an inspection of the whole line in August that year, the official opening of the West Cornwall Railway took place on 25 August 1852 to much local celebration. The first train was hauled by three locomotives carrying hundreds of guests crammed into forty wagons, arriving safely but late in Penzance, to be greeted by yet more crowds.

When completed, the line from Penzance to Truro was a combination of old and new and a world away from the grand designs of Brunel's Great Western Railway. The route transferred from new sections of line to older parts of the Hayle Railway between Hayle and Camborne and when the railway was inspected, the poor quality of the old permanent way was highlighted as a matter of concern. The new line did, however, feature some small but well-designed wooden station buildings in Brunelian style, similar to those seen on the SWR and OW&WR. The route chosen by the engineer meant that there was only one tunnel at Redruth and even that was only forty-five yards long. Further economies were made by the construction of nine timber viaducts reputed to have cost around £4,000 each, compared to around £20,000 for a stone-built structure. While cheap, the viaducts were safe, but many felt more than a little trepidation crossing them; 'I have great confidence in Mr Brunel as an engineer,' a WCR Director noted, but adding that he found some of his bridges 'rather frightful' due to their 'height and narrowness!'[46] However ramshackle some sections of line might have seemed, these economies did mean that the final cost of the line was nearer £100,000 than the £292,000 originally estimated by Brunel back in 1848, a result no doubt welcomed by the West Cornwall Directors.

The opening of the West Cornwall was widely welcomed in the county and sentiments expressed by the Town Clerk of Penzance during the opening celebrations would have been shared by many, including shareholders of the larger Cornwall Railway. The completion of the West Cornwall meant, he noted, 'the introduction of a line of railway through Cornwall…the first successful effort for breaking down the barrier which has, for so long a time, separated this country (sic) from the eastern parts of England'.[47]

By August 1852, some tentative steps had been taken to revive the fortunes of the Cornwall Railway; the year before, Brunel had suggested to Directors that by building a single line only, as he had done on the WCR, he could find contractors who would be able to construct the railway from Plymouth to Truro including stations and the bridge over the Tamar for around £800,000. Despite the best efforts of the company, money was still short, and there was only enough to employ contractors Sharpe & Sons to begin work on the fifteen-mile Truro to St Austell section in 1852.

By the end of that year, further contracts had been let for more sections of line including the railway from Devonport to the Saltash bridge and a contract for the construction of the bridge itself. Brunel worked to improve the cashflow of the struggling company by persuading local property owners to accept shares in the railway as payment for land, his long experience of dealing with the country gentry proving helpful in often delicate negotiations. Not all landowners were charmed by his approaches, however: Lord Vivian who owned land in the Glynn valley described him as 'a shifty personage, without any sense of fair dealing'.[48] By the summer of 1853 Brunel reported that good progress was being made, but that labour was in short supply due to the remoteness and rural economy of Cornwall. He suggested that one of the contractors 'should not commence work till after the harvest' and start with tunnelling, then 'limited and increasing quantity of earthworks and masonry during the next twelve months'. This arrangement should mean that 'the serious consequence of a sudden demand for labour in a rather remote district may be avoided.'[49]

In August 1853 Brunel was able to report to Directors that he was 'steadily progressing in letting the remainder of the works and executing those already contracted for'. He added that other works would begin at times 'so as to secure their completion at the proper time without needlessly advancing the expenditure of capital', indicating that funding was still proving an issue for the company. Contracts had also been let for the building of a branch from Truro to Falmouth, but with funding still difficult, work on this section was abandoned and not restarted for some years. With shareholders reluctant to support the railway company, Directors were forced to ask the Associated Companies to help, and it was agreed that they would lease the line for seven years after opening. This enabled work to begin on the Liskeard-Saltash section, but even this injection of funds was not enough and in March 1857 a further Act of Parliament was passed extending the time needed to complete the railway by another three years.

The railway would eventually take thirteen years to complete, with Brunel's attention split after 1852 between the more familiar task of managing the construction of the line and the separate but parallel project to build the Saltash Bridge, a task which would occupy a good deal of his time. For many engineers, this might be more than enough to cope with, but it should also be remembered that at the same time Brunel was engaged in the prolonged struggle to build the largest ship in the world, the SS *Great Eastern*, a task that would ruin his already fragile health and hasten his early death in 1859.[50] By August 1855, Brunel told Directors that 'Upon the whole everything has as yet proved entirely successful, the principal difficulties have been overcome and without hoping to avoid all casualties or course of delay the progress of the work, if slow may be considered sure.' He recorded that much of the work was in a forward state and that the line between Truro and St. Austell was 'with trifling exceptions completed and ready to receive the permanent way'.[51]

The Cornwall Railway provided the most perfect display of Brunel's timber viaducts and bridges; traversing a landscape of deep river valleys and tidal creeks, the line featured no fewer than thirty-four viaducts between Plymouth and Truro and another eight on the Falmouth Branch. Brunel by now had almost twenty years of experience of using timber for bridges and viaducts and had experimented not only with their design but also the different types of wood available and its preservation. This experimentation came to fruition in Cornwall where he built some of the largest and most impressive timber viaducts yet seen on his railways. While avoiding the cost of masonry viaducts was a matter of expediency on a railway always short of money, Brunel was also reluctant to use cast iron for bridges.[52] Some smaller viaducts were built in place of embankments where there was not enough spoil nearby to complete them. In the end, the relatively low cost of building timber viaducts enabled the railway to be built within the modest budget then available; Brunel told the Cornwall Railway board that the widespread use of timber would after ten years cost the company upwards of £10,000 per year in maintenance, a warning that would be proved right on that railway and many other lines he engineered. The GWR would inherit the problem and it would not be until the 1930s that the last timber viaduct had been expensively replaced by brick or masonry.[53]

Brunel's reports to Cornwall Railway Directors provided regular updates on progress on the largest structure on the line, the bridge at Saltash. Brunel's original plan for spanning the River Tamar had been for a massive timber structure with a central span of 255 feet with six other spans each of 105 feet. This idea was rejected when the Admiralty insisted that at high tide there should be headroom of at least 100 feet for any vessel passing under the bridge. Even more dramatic would have been an alternative idea involving the construction of a single span bridge of over 1,000 feet, removing the need for a central pier, but this proved unrealistic given that it would have cost more than £500,000, a sum the Cornwall Railway could clearly not afford. Instead, Brunel produced a design for a bridge that consisted of two main spans, each 455 feet long, supported by a central pier.

Considerable preparation and planning took place before construction began; a survey was undertaken to find the best place to build the foundations for the central pier, and as early as 1849 Brunel designed a tube that could be sunk into the thick mud of the riverbed allowing rock samples to be collected. Little else was done on site until 1853, but in the meantime Brunel had been developing ideas about the use of wrought iron for larger structures at Chepstow and on a smaller bridge built at Bristol Docks. The decision to make the railway single track reduced the cost of the proposed bridge at Saltash and further savings were achieved when Brunel negotiated with the contractor C. J. Mare, who agreed to build the bridge for just £162,000. The building of the central pier provided a huge challenge; to reach the bed rock, it was necessary to excavate almost eighty feet below the high water mark; this was done using the 'Great Cylinder', an even larger tube built on the river bank and then floated out into position in May 1854. Reporting the success of this operation, Brunel noted its position was within four-tenths of an inch of where intended, but that 'this extreme accuracy is of course accidental as none of our means of ascertaining the position or fixing it during the operation admitted of such accuracy.'[54] It was allowed to sink slowly into position and with the water pumped out, stonemasons were able to work in what was effectively a diving bell, Brunel drawing on his early experiences at the Thames Tunnel. Conditions within the cylinder were tough for the workmen, who battled both the water that was constantly pumped out and the effects of working for prolonged periods under extreme air pressure.

By the end of 1856 the great granite pier had finally been completed and the tube was dismantled and floated back to shore. The whole bridge was to be over 2,200 feet long, with seventeen side spans varying from seventy to ninety feet in length curving out across the river. Two large masonry towers were situated at each end and the two main spans were then carried by the central pier, which consisted of four enormous cast-iron octagonal columns. Brunel's design for the bridge trusses consisted of two arched wrought-iron tubes, oval in cross section, each 455 feet long. From each of the trusses hung suspension chains, connected by eleven uprights to which a wrought iron girder carrying the track was attached, with diagonal bracing added for further rigidity.[55]

At Saltash Passage on the Devon side of the Tamar, a yard described as 'a place of bustle and business' contained workshops where the construction of the bridge trusses took place. When the contractor C. J. Mare was declared bankrupt in 1853, Brunel took personal charge of the project, working closely with a team of assistants including Bertram and R. P. Brereton. Brereton had begun his career as a pupil of Brunel and had worked with him as assistant and resident engineer on the GWR, Taff Vale, C&GWUR, South Wales and Bristol & Gloucester lines. In later years, as Brunel's responsibilities had increased and his health had diminished, Brereton had effectively become Brunel's head of engineering staff and had been closely involved on the construction of the Chepstow Bridge.[56] The task of construction of the trusses was clearly complex and Brunel wrote to Charles Saunders in August 1854 to tell him that 'Brereton, Bertram and I have been at work incessantly – this bridge affair having involved a serious amount of labour,' noting that ensuring that each of the thousands of wrought-iron plates needed for each truss was uniform in size and thickness was a challenge. Admitting that he had not realised the amount of work involved, he conceded that Brereton was doing much of the work 'and indeed without him it would be impossible'.[57]

By September 1857 the first truss was complete. Weighing more than 1,000 tons each, lifting them into position would not be easy, but Brunel already had experience gained at Chepstow and had assisted Robert Stephenson at the lifting of the Conway and Britannia bridges. There had been time to plan and test arrangements before operations began, a luxury not enjoyed by Brunel the following year for the launch of the SS *Great Eastern* and he also had the support of old friend Christopher Claxton and Captain Harrison, Master of the SS *Great Eastern* for 'arrangements afloat'. Brunel, realising the complexity of the operation around the river instructed staff on 'the necessity of paying great attention and shewing great civility on all occasions to the various persons in authority that you may come into contact with…we shall want the good will of fishermen, lightermen, steamboat men – indeed everybody.'[58] On 1 September 1857 with an estimated 300,000 people watching from both sides of the river, the first truss was winched onto pontoons and floated out into the river. Brunel was in charge of operations and the movement of the pontoons and winches was controlled by a series of flag signals. Brunel called for complete silence during the operation to prevent any confusion and the trusses were carefully positioned on the base of the bridge piers. Rather than build the whole of the supporting bridge pier, Brunel instead decided to build it up under the truss. Three hydraulic jacks at each end lifted the thousand-ton truss up at three feet intervals while workmen built masonry below.

By May 1858, the first truss had been lifted 100 feet up into its correct position, and on 10 July the second was floated out into position in a similar operation, this time directed by Brereton, as Brunel was away in Europe convalescing. The second event was marked by more public celebration, and after the floating out, an 'excellent luncheon' was held for guests including Directors of the GWR, Bristol & Exeter and Cornwall railways, as well as local dignitaries. Brereton sent a telegram to Brunel, then resting in Lausanne, to let him know all had gone well.[59] The second truss was complete by December 1858 and it was Brereton who was left to manage the final stages of bridge construction as Brunel's health continued to worsen; he and his family left for Egypt in December that year, spending Christmas in Cairo with Stephenson and returning to England in the spring of 1859. A report written to Directors in February 1859, probably compiled by Brereton or one of Brunel's other resident engineers, noted that the line from Plymouth to Saltash had been completed and that the Saltash bridge was about be tested by the running of heavy trains across it.

Brunel was not well enough to attend the ceremony held on 2 May 1859 when the bridge was officially opened by Prince Albert, who had agreed to the bridge bearing his name in 1853. The Prince Consort, accompanied by Daniel Gooch, had left Windsor at 6.00am, travelling the whole length of Brunel's broad gauge line to the West Country, arriving at Saltash at around noon. After the usual speeches of welcome,[60] the royal party travelled across the bridge by train and then the Prince walked back across the structure with Brereton. A 'cold collation' followed, after which the royal party left for a river cruise, returning to Windsor by train at 7.00pm. Brunel was not able to see his creation until later in the month when he made a private visit and crossed his bridge lying on a couch placed specially on a railway wagon, a poignant image of a man whose health had finally been broken by his other great work, the steamship *Great Eastern*.[61] The railway opened for passengers on 4 May 1859 and in August Directors were told by Brunel that 'since the opening of the Railway the Works and Permanent Way have stood well and the Viaducts are in a very satisfactory condition.'[62] The engineer would not live to see the completion of the Falmouth Branch in 1863, or the final extension of the broad gauge to Penzance via the West Cornwall Railway in 1867, completing the railway network he had begun to envisage when he wrote his speculative diary entry on Boxing Night 1835.

EPILOGUE

After Brunel

Years of long hours, lack of sleep, anxiety and a fearsome appetite for cigars had taken their toll on Brunel's health, although the traumatic process of supervising the construction and launch of the *Great Eastern* steamship at Millwall clearly hastened the onset of an illness that would prove to be terminal. In November 1858 he had been diagnosed with Bright's Disease, a severe condition affecting the kidneys and was once again sent away to rest in warmer climes. When he returned to England in May 1859, it was remarked that he looked better, but his remission was short. With the Royal Albert Bridge and Cornwall Railway complete, Brunel spent much of his time visiting his great ship, now being fitted out on the Thames. It had been arranged that he would travel on the SS *Great Eastern's* first trip to Weymouth on 7 September, but it was not to be. Two days before the voyage Brunel suffered a stroke after making an inspection of the ship and was taken back to Duke Street. There he remained receiving daily reports about the ship, dictating letters and putting his affairs in order and despite rallying a little, on the afternoon of 15 September 1859 he called his family together and after speaking with them passed away peacefully later that evening.

As well as tributes paid to Brunel in local and national newspapers, his work was also acknowledged by many of the railways he had been involved with; speaking at the first half-yearly meeting of the Cornwall Railway held after his death, the Chairman Dr G. Smith told shareholders that the company owed a great debt to Brunel, 'whose works of extraordinary genius had earned him a European reputation' and that his loss would be not only to the company, but to 'the general interests of science throughout the world'. In a more parochial aside the Chairman also reminded them that the completion of the Saltash Bridge for £225,000 'could not have been more economically completed'.[1] It was shortly afterwards that Directors agreed to inscribe Brunel's name on the towers of the Saltash Bridge as a permanent reminder of his contribution to both their railway and the engineering world more generally.

With their master gone, it was left to staff and in particular his chief clerk Joseph Bennett to pass on the sombre news of his passing to clients and to wind up his affairs, a process that took a number of months. Sixteen circular letters were sent to the secretaries of railway companies with whom Brunel had been dealing, along with other individuals and concerns. Writing to the Swansea Harbour Company, for whom Brunel had been designing coal drops, William Barber stated that 'with deep regret, I have to convey the sad intelligence of the death of Mr Brunel…an unfortunate change took place last Wednesday afternoon and he died the same night at 25 minutes past 10 o'clock.'

Brereton wrote a further letter to railway companies at the end of September informing them that having been for twenty-three years in Brunel's service and 'for some time chief of his Engineering Staff, it has been determined that I should, if it meets the views of your company, carry on the works'. He concluded noting that Brunel had expressed a strong desire that his second son should follow him in the profession of a Civil Engineer, and 'it has been arranged… that ultimately when qualified he may have the opportunity of joining me.'[2] Brereton took responsibility for the completion of the West Somerset Railway, a project begun in 1856 when a company was set up to build a railway from Watchet to connect with the Bristol & Exeter Railway. Brunel 'had much to say' at the first public meeting in July that year and he suggested a route via Williton to Taunton, which avoided the need for an expensive tunnel. Although he undertook some initial survey work, it was Brereton who was appointed engineer in April 1859, assisted by a resident engineer, and the fourteen-mile railway was eventually opened in March 1862.

Another project completed after Brunel's death was the construction of a dock at Briton Ferry and the associated South Wales Mineral Railway. Demand for a new port to serve the iron, steel and tinplate works and other factories situated on the banks of the River Neath had led to the formation of a company in 1846, but it was not until the early 1850s that the project came to fruition. Brunel was also the engineer of the South Wales and Vale of Neath lines, so it made sense to connect the various railways with the dock; he was therefore appointed as engineer to both the Briton Ferry Dock Company and the South Wales Mineral Railway in 1853. The broad gauge line was a mineral railway that linked Briton Ferry with collieries in the Rhondda and included a 1 ½ mile incline that carried both locomotives and wagons up and down. Although construction work began in 1856 and it opened partly in 1861, the line was not finally completed until 1863. Brunel did not see the completion of his dock at Briton Ferry either; it opened in August 1861 to much celebration.[3] The local MP Mr H. H. Vivian told a newspaper that although the engineer had been 'removed from us, his works remain and will still do so. I believe Mr Brunel to be the greatest engineer who ever lived.'[4]

One of the final projects begun under Brunel and completed after his death was the Bristol & South Wales Junction Railway. The company was authorised by Act of Parliament in 1846 to build a railway from Bristol Temple Meads to the banks of the River Severn at New Passage. The line would connect with a ferry to Portskewett in South Wales, where a short branch would connect with Brunel's South Wales Railway; Brunel spent time working up proposals but the scheme was eventually abandoned for lack of money. In 1854, a new railway was proposed, the Bristol, South Wales and Southampton Union Railway.[5] Plans were finally approved by Parliament in 1857, with Brunel using much of the work he had done previously. He wrote to his old colleague Charles Richardson, whom he had once rebuked for playing too much cricket instead of working, now using the attraction of cricket to tempt him into working for him again. He told him that he needed 'a man acquainted with tunnelling...who will look after the tunnel with his own eyes'. The country immediately north of Bristol 'I should think a delightful one to live in,' he added, 'I can't vouch for cricket but I think it highly probable.'[6] Richardson accepted on a salary of £300 per year and work began on the railway. It proved to be a slow process, and the line did not finally open until 1863 when trains ran from Bristol to a large wooden pier designed by Brunel that survived until 1887.

APPENDIX 1

Brunel's Railways

The following list tries to capture the numerous railway projects Brunel was directly involved with as an engineer in the order they opened, with the aim of illustrating the chronological, commercial and geographical spread of his work. Where this process took place over a number of years, as in the case of a company such as the Wilts, Somerset & Weymouth, multiple dates have been included. Not listed are the various projects for which Brunel undertook only surveys, or acted as consultant and which were not constructed. This is not a definitive list but includes the most significant lines he was engaged with, often simultaneously. Irish railways are included here but Brunel's work overseas is noted in Appendix 2.

Railway Company	**Opening Dates**
Great Western Railway	1838–1841[1]
Taff Vale Railway	1840–1841
Bristol & Exeter Railway	1841–1844
Bristol & Gloucester Railway	1844
Cheltenham & Great Western Union Railway	1845
South Devon Railway	1846–1848
Berks & Hants Railway	1847–1848
South Wales Railway	1850–1856
Wilts, Somerset & Weymouth Railway	1850–1857[2]
Oxford & Rugby Railway	1850–1852
Gloucester & Dean Forest Railway	1851

Vale of Neath Railway	1851–1853
West Cornwall Railway	1852
Birmingham & Oxford Railway	1852
Oxford, Worcester & Worcester Railway	1852
Cork & Waterford Railway	1853
Hereford, Ross & Monmouth Railway	1853–1855
South Devon & Tavistock Railway	1856
Waterford, Wexford, Wicklow & Dublin Railway	1856
East Somerset Railway	1858
Cornwall Railway	1859
South Devon & Tavistock Railway	1859
Dartmouth & Torbay Railway	1859
Llynvi Valley and South Wales Junction Railway	1861
South Wales Mineral Railway	1861–1863
West Somerset Railway	1862
Bristol & South Wales Union Railway	1863

APPENDIX 2

Overseas Railway Projects

In common with other contemporaries like Locke and Stephenson, Brunel's abilities and public image meant that he was frequently approached by the promoters of numerous railway schemes overseas. As work on his railways began to develop significantly in the 1840s and the main line routes in Britain were established, the construction of railways in Europe, the United States and parts of the British Empire had also begun. Since Britain had been the cradle of railways it was therefore not unexpected that British engineering knowledge and experience, both in the surveying and construction of lines and the manufacture of locomotives and rolling stock, would be in great demand.

There is little doubt that Brunel's prodigious workload that encompassed not only the railways described in this volume but also steamships, bridges and other projects meant that he undertook far fewer railway commissions abroad than other railway engineers who in the 1850s were heavily involved in work overseas. Given the cosmopolitan nature of his upbringing and his family connections with France, Brunel seems to have spent relatively little time abroad until the last years of his life, when he travelled to Europe and Egypt to convalesce. He did not undertake long voyages on any of his steamships and largely remained resolutely at home, although as his diaries record, he travelled extensively around Britain in the course of his work.

The frenetic pace at which Brunel worked in the 1830s would have left little time for foreign travel and he reflected in 1841 that he had 'hitherto invariably declined any foreign engagements for which I have had many proposals'[1] on account of his work at home. In addition to surveying and supervising the construction of the GWR, Bristol & Exeter, Bristol & Gloucester, Cheltenham & Great Western Union and Taff Vale railways he had also found time to design the SS *Great Western* and work intermittently on his first love, the Clifton Bridge. Not surprisingly then, it was not until the following decade that he felt ready to undertake work away from Britain. This is not to say that his workload had diminished substantially, with the construction and launch of the SS *Great Britain* occupying much time along with challenging projects like the atmospheric railway in Devon.

Brunel undertook a number of railway projects in Italy and was first approached at the end of 1841 with a proposal from the government of Piedmont to survey a line from Turin to Genoa. As usual, Brunel was insistent that he would be considered as 'principal engineer and first authority on all engineering matters'; he undertook to do the work for a fee of ten guineas per day plus expenses, stating he would only stay in hotels when visiting Italy.[2] The main work of the survey was done by experienced surveyor William Johnson assisted by Benjamin Babbage, son of his friend Charles Babbage, the mathematician. A report was submitted to the government and after a long delay the project finally seemed likely to start in 1844. Brunel and his wife Mary had visited Italy that summer to inspect the route, but the project was abandoned in November that year without any work being done.

A further opportunity presented itself in 1845 when Brunel was asked to build the Maria Antonia Railway from Florence to Pistoia. Babbage was once again his man on the ground and while the railway was completed by 1848, Brunel was not fully paid for his services, despite numerous attempts to extract payment. Italy was in the midst of revolution again and the government of the Duchy of Tuscany probably was preoccupied with other matters at the time.

In the 1850s Brunel received other offers from railway promoters across the globe. Records reveal proposals for other lines in Italy such as Rome to Ancona and Alessandria to Piancenza, but these do not seem to have been proceeded with. In addition, Brunel was asked for assistance with railway projects across the Atlantic in both the United States and Canada. His appointment diaries provide tantalising insights into the range of work he was asked to comment on: on 24 March 1851 for example, at the end of a busy day it was recorded that 'at 8:20pm Mr Brunel met with Mr Home regarding the Halifax and Quebec railway.'[3]

After 1855 Brunel became involved in plans for the East Bengal Railway, a 150-mile railway that would run north from Calcutta. His increasing involvement in the construction of the SS *Great Eastern* and his declining health meant that visiting India was impossible, and work was entrusted to the engineer W. A. Purdon. The construction of the railway was delayed by the rebellion, or the First War of Indian Independence, and not completed until after Brunel's death.[4] A Brunel connection to the railway survived for some time since his assistant R. P. Brereton attended the East Bengal board in his place, and later his son Henry was also involved with the company.

Brunel's final overseas project was work for the Melbourne to Williamstown Railway in Victoria, Australia, in the final years of his life. A number of contracts were delivered for the Government of Victoria for bridges and other works along the line including a substantial wrought-iron bridge across the Saltwater Creek designed in Brunel's Duke Street office. This remains the only tangible relic of his work in Australia, although the influence of the thousands of passengers who arrived in Melbourne having travelled halfway around the world on his pioneering steamship the SS *Great Britain* is arguably more important to the history of that place.

APPENDIX 3

Timeline

1769	Birth of Marc Isambard Brunel.
1799	Marc Brunel arrives in England after exile in America and marries Sophia Kingdom.
1806	9 April - Isambard Kingdom Brunel born at Portsea, Portsmouth.
1820	Isambard Kingdom Brunel sent by his father to school in France aged fourteen.
1821	Marc Brunel imprisoned for debt.
1822	Brunel Returns to England to join father's drawing office in London.
1825	Work begins on Thames Tunnel.
1827	Isambard appointed Resident Engineer on tunnel when his father falls ill. The tunnel floods for first time.
1828	Brunel is seriously injured in second flood at Thames Tunnel in January. Brunel learns of plans to build bridge across the Avon at Clifton while convalescing in Bristol.
1829	Brunel submits designs for first Clifton Bridge competition, but they are rejected by Thomas Telford.
1831	Winning designs for second Clifton Bridge accepted. Construction begins on bridge, but Bristol Riots stop work in September. Brunel takes first trip by train on the Liverpool & Manchester Railway.
1833	Brunel appointed as engineer of the Great Western Railway and completes survey of the route.

1834	First Great Western Railway Bill rejected in Parliament.
1835	Second Great Western Railway Bill passed in Parliament. GWR Directors accept Brunel's broad gauge proposal. Brunel appointed as engineer for Cheltenham & Great Western Union, Bristol & Exeter, Bristol & Gloucester, and Merthyr & Cardiff railways. Brunel proposes building a steamship for the Bristol-New York service.
1836	Work begins on construction of the GWR. Brunel appointed engineer of Great Western Steamship Company. Brunel marries Mary Horsley and moves to Duke Street Westminster.
1837	Daniel Gooch appointed as Locomotive Superintendent to the GWR. SS *Great Western* launched. *North Star* locomotive delivered to the GWR.
1838	First section of GWR from Paddington to Maidenhead opened. SS *Great Western* makes its maiden voyage to New York. Brunel's first son Isambard born.
1839	Maidenhead Bridge opened and GWR line extended to Twyford. Appointed engineer to Bristol & Exeter Railway. Brunel survives challenge from Liverpool Party shareholders over broad gauge. Construction of SS *Great Britain* begins in Bristol.
1840	Opening of GWR from Reading to Hay Lane, and from Bristol to Bath. Taff Vale Railway opened from Cardiff to Merthyr. GWR Board approves Swindon as site for the company's locomotive works.
1841	Box Tunnel completed and Bristol Temple Meads station opened. GWR line from London to Bristol opened throughout. Appointed engineer of Oxford Railway.
1842	Opening of Bristol to Taunton section of Bristol & Exeter Railway. Birth of second son Henry Marc Brunel.
1843	Swindon locomotive works opened. Oxford Branch completed. Appointed engineer of South Devon Railway. Launch of SS *Great Britain* at Bristol. Brunel almost dies when he swallows a half-sovereign.

1844 Bristol & Exeter Railway opened throughout to Exeter. Appointed engineer to Berks & Hants, Monmouth & Hereford, Oxford, Worcester & Wolverhampton, Oxford & Rugby, South Wales, Wilts, Somerset & Weymouth railways.

1845 Work begins on South Devon Railway. Cheltenham & Great Western Union Railway opens. Battle of the Gauges rages and Gauge Commission set up by Parliament. Appointed engineer of Cornwall and West Cornwall railways. Maiden voyage of SS *Great Britain*.

1846 The Gauge Commission rules in favour of the 'standard gauge' and Gauge Act passed preventing new broad gauge lines being built. South Devon Railway extended to Teignmouth and Newton Abbot. SS *Great Britain* runs aground at Dundrum Bay.

1847 Atmospheric system used on Exeter-Teignmouth section of South Devon Railway. Birth of Florence, Brunel's daughter. 17 Duke Street purchased.

1848 Atmospheric system extended to Newton Abbot on South Devon Railway,

1849 South Devon Railway completed to Plymouth but atmospheric system is abandoned. Brunel begins design of Paddington station. Death of Sir Marc Isambard Brunel.

1850 South Wales Railway opens from Chepstow to Swansea.

1851 First section of Vale of Neath Railway completed.

1852 Work begins on construction of new Paddington station. West Cornwall Railway opened from Truro to Penzance. Brunel resigns as engineer of the Oxford, Worcester & Wolverhampton Railway. Appointed engineer of Eastern Steam Navigation Company to build the *Great Eastern* steamship.

1853 Chepstow bridge completed and South Wales Railway linked to GWR network. Brunel appointed engineer of Briton Ferry dock project.

1854	Paddington Station fully opened. Wycombe Railway completed. Work begins on construction of SS *Great Eastern* at Millwall.
1855	Opening of Hereford, Ross & Monmouth Railway.
1857	First span of Royal Albert Bridge floated into position. Final section of the Wilts, Somerset & Weymouth Railway completed. SS *Great Western* broken up. First attempt to launch SS *Great Eastern* fails.
1858	Second span of the Royal Albert Bridge successfully installed. SS *Great Eastern* launched. Brunel's health deteriorates and he travels abroad to convalesce.
1859	Royal Albert Bridge at Saltash opened. Sea trials of SS *Great Eastern*. Brunel passes away on 15 September.
1864	Opening of Clifton Suspension Bridge.
1892	The last broad gauge trains run on the GWR and final section converted to standard gauge in May.

Endnotes

BI	Brunel Institute
GWT	Great Western Trust, Didcot
STEAM	STEAM: Museum of the Great Western Railway, Swindon
TNA	The National Archives

Introduction

1. Quoted in: Burgess, D.R., *Engines of Empire* (Stanford University Press, 2016) p 89.
2. Samuel Smiles, the nineteenth-century chronicler of railway engineers noted that 'Mr Brunel always had an aversion to follow any man's lead; and that another engineer had fixed the gauge of a railway, or built a bridge, or designed an engine in one way, was of itself often a reason for him adopting an altogether different course.'
3. Latimer, T. *Annals of Bristol in the Nineteenth Century* (Bristol: W.F. Morgan, 1887) p 191.
4. Vaughan, A *Isambard Kingdom Brunel: Engineering Knight-Errant* (London: John Murray, 1991) p 271.
5. Burgess, D.R. p89.
6. BI: DM1306/2/1/folio 3 & DM1306/2/3/2 folio 169.
7. BI: DM1306/7/5 I.K Brunel to Thomas Guppy 11 May 1838.
8. BI: Copy of 'Sketch of Mr Brunel' *Manchester Guardian*: reprinted in the *Goulburn Herald & County of Argyle Advertiser* 13 February 1858. The correspondent was describing Brunel's appearance during the abortive launch of the SS *Great Eastern* but there is little doubt that his apparel that day was his normal dress style.

Chapter 1

1. BI: DM1306/2/1 I.K Brunel Diary 1835.
2. See also Angus Buchanan's *Brunel's Locked Diary* (Bristol: BIAS, 1996) which includes a transcription of the diaries and further commentary. The diary is displayed in the 'Being Brunel' Museum at the SS *Great Britain.*
3. BI: I.K. Brunel Letter to Benjamin Hawes 27 March 1831.
4. Chrimes, M. 'Castles in the Air to the Bristol Railway. How did Isambard Kingdom Brunel make his living 1828-1833?' *Panel for Historical Engineering Works Newsletter* No. 109 March 2006 pp 1-2.
5. For more detail on the story of the Clifton Suspension Bridge see: Andrews, A. & Pascoe M. *Clifton Suspension Bridge* (Bristol: Broadcast Books, 2008).
6. Hill, R. *God's Architect: Pugin and the Building of Romantic Britain* (London: Penguin, 2008) p 89.
7. Quoted in: Andrews, A. & Pascoe M. *Clifton Suspension Bridge* (Bristol: Broadcast Books, 2008) p 24.
8. Webb, W.W. *A Complete Account of the Clifton Bridge* Bristol:1865 pp 23-24.
9. For more detail on the Bristol Riots see: Cannadine, D. *Victorious Century* (London: Penguin 2017) pp 153-166 and Latimer, T. *Annals of Bristol in The Nineteenth Century* (Bristol: W & F Morgan, 1887) pp 146-177.
10. Brunel described the committee as 'a set of deuced clever fellows – but a rum set'. For more detail of the dock works at Monkwearmouth see: Brunel, I. *The Life of Isambard Kingdom Brunel Civil Engineer*, 1870 (Reprinted edition: David & Charles 1971) pp 417-422.
11. See: Malpass, P & King, A. *Bristol's Floating Harbour: the First 200 Years* (Bristol: Redcliffe Press, 2009) pp 47-53.
12. TNA: RAIL1014/1 Extracts from the minutes of the Meeting of the Kennet & Avon Canal Company 19 January 1825.
13. TNA RAIL1014/1 'Proposed Rail-Road from Bristol to Bath' Bristol December 1 1824.
14. Rev. S Eyer, 'Chronicle of Events 1820–1827', quoted in: Gibson, C. *Bristol's Merchants and the Great Western Railway* (Bristol: Historical Association (Bristol Branch) 2002).
15. Hayes, D. *The First Railways* (London: Harper Collins, 2017) p 172.
16. TNA RAIL1014/1 'Bristol to London Railway' 7 May 1832.
17. TNA RAIL1014/1 'London to Bristol Railway' 13 June 1832.
18. Latimer, J. *The Annals of Bristol in the Nineteenth Century* (Bristol: W & F Morgan, 1887) p 189.
19. The committee consisted of Bristol Corporation: John Cave, Charles Ludlow Walker & John Lunnell. Society of Merchant Venturers: George

Gibbs, Peter Maze and Henry Fowler. Bristol Dock Company: Humphrey Jefferies, John Howell and Nicholas Roch. Bristol Chamber of Commerce: William Jacques, Edward Harley and William Tothill. Bristol & Gloucester Railway: John Harford, George Jones and Joseph Fry.

20. GWR: Printed Circular and letter of instruction dated 21 January 1834. Author's collection copy.
21. BI: Transcript of Brunel Diary for 21 February 1833 (Adrian Vaughan Collection).
22. Priestley, J. *Historical Account of the Navigable Rivers, Canals and Railways Throughout Great Britain* (London: Longman, 1831) pp 107-109.
23. BI: DM1306 III.1 GWR Handbill 1833.
24. Buchanan, R.A. *Science and Engineering: A Case Study in British Experience in the Mid-Nineteenth Century*. Reprint: *Notes and Records of the Royal Society of London* Vol 32, No.2 March 1978.
25. See also: Pendred, L. St. L. *British Engineering Societies* (London: Longmans Green & Co, 1947)
26. Tredgold, T. *The steam engine – its invention and progressive improvement, an investigation of its principles, and its application to navigation, manufactures, and railways* (London, J. Weale, 1827)
27. BI: Clive Richards Brunel Collection 2014.03135.
28. TNA: RAIL1008/74.
29. Long, P.J & Awdry, W.V *The Birmingham and Gloucester Railway* (Stroud: Alan Sutton, 1987 pp 4-5).
30. BI: DM/1306/2/3/1/folio 204 5 December 1831.
31. BI: DM1758-6-6 Brunel Pocket notebook 1838.
32. TNA: RAIL1014/1.
33. A sum equivalent to more than £190 million today.
34. BI: DM162/24 Calculation Book: Bristol Railway 1833-1834.
35. For more detail see: Gren, A. *The Foundations of Brunel's Great Western Railway* (Kettering: Silver Link Publishing, 2003) p 39.
36. For more information see: University College of London: *Centre for the Study of Legacies of British Slavery database* (https://www.ucl.ac.uk/lbs/)
37. Channon, G. *Bristol and the Promotion of the Great Western Railway* (Bristol: Bristol Branch of the Historical Association, 1985) p 8.
38. Great Western Railway Between Bristol and London: Prospectus, 1833 (Author's collection)
39. Benedict, E. 'Further Reminiscences of Isambard Kingdom Brunel' *Great Western Railway Magazine* March 1910.
40. See: Russell, J.H. *Great Western Miscellany* (Oxford: OPC, 1978) Plates 6 & 7.

Chapter 2

1. Parris, H. *Government and the Railways in Nineteenth Century* Britain (London: Routledge & Kegan Paul, 1965) p 18.
2. BI: Copy of Resolution of Bristol Committee 4 September 1833 (Adrian Vaughan Collection)
3. BI: Copy of Resolution of London Committee 7 September 1833 (Adrian Vaughan Collection)
4. BI: Copy of letter from Charles Saunders to I.K. Brunel 7 September 1833 (Adrian Vaughan Collection)
5. See: Conder, F.R. *The Men who Built Railways* (London: Thomas Telford Ltd. 1983)
6. TNA: RAIL 1008/69 W. Hughes to I.K. Brunel.
7. BI: DM1306 III.2 I.K. Brunel *Instructions to Surveyors* 1833.
8. George Eliot describes a violent encounter between railway surveyors and farm labourers in Chapter 56 of *Middlemarch* published in 1871 but set thirty years earlier.
9. Letter from Charles Saunders to Isambard Kingdom Brunel 28 September 1833. Senate House Library, University of London Ref: AL238.
10. BI: DM1758/2/1 & DM1758/2/2 Brunel Appointment Diary for 1834.
11. Parliamentary standing orders required that 50% of shares should be subscribed before a bill could be read for the second time in the house.
12. 'Account of Public Meeting Held at the Merchants Hall, Bristol 15 October 1834', *The Berkshire Chronicle* 18 October 1834.
13. BI: DM1758/2/1 & DM1758/2/2 - Brunel Appointment Diary for 1834.
14. See: Channon, G. *Bristol and the Promotion of the Great Western Railway* (Bristol: Bristol Branch of the Historical Association, 1985)
15. Berkshire Chronicle 17 March 1834.
16. TNA: RAIL 1014/1 *Extracts from the Minutes of Evidence given before the House of Commons on the Great Western Railway Bill* (Bristol: Gutch & Martin, 1834)
17. Vaughan, A. *Isambard Kingdom Brunel: Engineering Knight Errant* (London: John Murray, 1992) p 51.
18. Sekon, G.A. *A History of the Great Western Railway* (London: Digby Long & Co, 1895)
19. Quoted in: Phillips, D. *How the Great Western Came to Berkshire* (Reading: Berkshire County Library, 1985) p 7.
20. See: Francis, J. *A History of the English Railway: Its Social Relations & Revelations 1820–1845* (London: Longman Brown, Green & Longmans, 1851)

21. South, R. *Crown, College and Railways* (Buckingham: Barracuda Books, 1978) p 14.
22. TNA: RAIL 1014/1 *Report of Henry Habberley Price Civil Engineer, relative to a Great Western Railway Communication from London into South Wales...with a comparative view of the merits of the London and Reading and London and Windsor Railway Scheme.* 7 March 1834.
23. TNA 1014/1 *Reasons Why the Great Western Railway Bill Now Before Parliament Should Not Pass Into Law*, Handbill, March 1834.
24. Sekon, G.A. *A History of the Great Western Railway* p 6.
25. Gren pp 53-54
26. BI: DM1306 III.4 *IKB's notes on Mr Joy's summing up*.
27. Sekon, G.A. *A History of the Great Western Railway* p 5. Ipecacuanha was a herbal medicine used as an expectorant.
28. 'Great Western Railway: Report of a Public Meeting held on Wednesday 13 October 1834 at the Merchants Hall, Bristol'. *Berkshire Chronicle* 18 October 1834.
29. See: Channon, G. *Bristol and the Promotion of the Great Western Railway* (Bristol: Bristol Branch of the Historical Association, 1985)
30. TNA: RAIL1014/1 Great Western Railway Handbill, 47, Parliament Street London, 30 May 1835.
31. Russell could hardly be described as a neutral, since he had been a strong supporter of the railway for some time. He was its Chairman from 1839 to 1855.
32. TNA: RAIL 1014/1 *Petition of Coachmasters, Proprietors of Vans, Waggons, Postchaises and other Carriages and of Horses for Hire*, 1835.
33. Swift, A. *The Ringing Grooves of Change* (Bath: Akeman Press, 2006) p 39.
34. TNA: RAIL 1014/1 Bath & Basing Railway: Meeting at Frome, Wednesday 17 December 1834.
35. TNA: RAIL1014/1 *Reply to the Case of the Opponents of the Great Western Railway* (London: Savill's Printer, 1835)
36. Sekon, G.A. *A History of the Great Western Railway* (London: Digby Long & Co, 1895) pp 4-5.
37. Levien, D. V. 'A Hundred Years Ago: How the Great Western Came into Being', *Great Western Railway Magazine*, August 1835, p 443.
38. The first meeting in Bristol of the British Association was held in August 1836. Amongst those present were Professors Babbage, Faraday, Messrs. I. K. Brunel and Dr Lardner who 'was so unwise as to prove to his own satisfaction the impossibility of a steamboat ever crossing the Atlantic'. Nicholls, J. F. and John Taylor. *Bristol Past and Present, Vol. III – Civil and Modern History* (Bristol: J. W. Arrowsmith, 1882) p. 345.

39. BI: Great Western Railway Handbill: *printed letter from George Stephenson and Henry Palmer,* London 31 March 1835.
40. BI: DM1758/14.1.1 *Letter from George Stephenson & Henry Palmer to the Directors of the GWR* 31 March 1835.
41. BI: Transcript of Notes of 1835 Committee proceedings: Adrian Vaughan Collection.

Chapter 3

1. Clark, E.T. *The Great Western Railway Guide Book* (London: Smith & Ebbs, 1838) p 14.
2. TNA: RAIL1149/2 I.K. Brunel to George Frere, 3 September 1835.
3. Whishaw, Francis *The Railways of Great Britain & Ireland* (London: John Weale, 1842 David & Charles Reprint 1969) p 141.
4. As a comparison, Stephenson's London & Birmingham Railway, opened in 1830, had a ruling gradient of 1 in 330.
5. Clark, E.T. *The Great Western Railway Guide Book* p 14.
6. TNA: RAIL250/82 I.K. Brunel: Letter and report to GWR Directors 19 September 1835.
7. In the event few vehicles of this type were ever built or used on the GWR.
8. BI: DM1306 III.7 Draft agreement for shared terminus with London & Birmingham Railway at Euston 1834.
9. *The Great Western Railway Magazine: A Miscellany of Fact and Fiction* Vol II (London: George Burns & Co, 1864)
10. Brunel was eventually able to acquire 17 Duke Street in 1848.
11. See: R.A. Buchanan 'Working for the Chief' in *Isambard Kingdom Brunel: Recent Works* (London: Design Museum, 2000) p 25, for a list of assistants and engineers who worked for Brunel.
12. Letter quoted in Vaughan, A. *Isambard Kingdom Brunel: Engineering Knight-Errant* (London: John Murray, 1991) p 63.
13. Quoted in Rolt, L.T.C. *Isambard Kingdom Brunel* (London: Longmans, 1957) p 107.
14. BI: DM1298/2/38 & DM 1298/2/71 letters from J.W. Hammond to R. Archibald, 14 April and 14 July 1837.
15. Journal of Charles Richardson 1835–1838 entries for 1836. (Courtesy of Mr Howard Beard). Richardson would go on to have a distinguished career as a railway engineer, most notably working with Charles Walker on the construction of the Severn Tunnel in 1886.
16. R.A. Buchanan 'Working for the Chief' in *Isambard Kingdom Brunel: Recent Works,* p 16.
17. BI:DM1758/5/5 Appointment Diary for 1843.
18. BI: DM1758/5/3 Brunel Appointment Diary for 1836.

19. TNA: RAIL 250/1 Great Western Railway: Report of the Second Half-yearly General Meeting 25 August 1836.
20. TNA: RAIL1110/452 Handbill of meeting held on 1 October 1835 at Merthyr Tydfil to form the Taff Vale Railway.
21. BI: DM162/8/2/6/folio 15 Notes about Merthyr and Cardiff railway 2 November 1836.
22. Quoted in: Jones, S.K. *Brunel in South Wales: Volume 1 In Trevithick's Tracks* (Stroud: History Press, 2005) p 124.
23. For further detail on the construction of the Taff Vale Railway see: Jones, S.K. *Brunel in South Wales: Volume 1.*
24. Noble, C. *The Brunels: Father & Son* (London: Cobden-Sanderson, 1938) p 124.
25. Brunel, I. *Life of Brunel* 1870 pp 233-241.
26. 'Opening of the Great Western Railway', *The Times* 2 June 1838. Guppy, speaking at the dinner following the opening of the GWR from Paddington to Maidenhead, may have misremembered events since he told guests that the dinner had been held during the committee stages of the GWR bill. This could not have been the case, as the bill had been passed two months previously in August 1835.
27. Rolt, L.T.C. *Isambard Kingdom Brunel* p 105.
28. For more information about the National Brunel collection held by the SS Great Britain Trust see: https://www.ssgreatbritain.org/collections-and-research/collections/
29. Many of the most significant engineering drawings relating to Brunel's Great Western Railway have been digitised by Network Rail and can be found at: https://history.networkrail.co.uk/
30. Clifford, D. *Isambard Kingdom Brunel: The Construction of the Great Western Railway* (Reading: Finial Publishing, 2006) pp 47-48.
31. Conder, F.R. *Personal Recollections of English Engineers* (London: Hodder & Stoughton, 1868) p 244.
32. Charles Richardson Diary 1835–1838.
33. 'TO BUILDERS AND OTHERS: Request for tenders for construction of Reading Station', *Windsor & Eton Express* 22 June 1839.
34. GWT: Tender for contract 4R. J. Bedborough 22 March 1838.
35. Conder, F.R. *Personal Recollections of English Engineers* p243.
36. BI: DM.1758/15 I.K. Brunel to Charles Saunders 3 December 1837.
37. TNA: RAIL 250/105 Meeting of the Great Western Railway Bristol Progress sub-committee 17 July 1838.
38. TNA: RAIL1149/2 I.K. Brunel to George Frere. 3 September 1835.
39. See: *Great Western Railway Magazine* March 1906 p 102. Grissell & Peto's skew bridge had a very short life, destroyed by fire in 1847, being replaced with a Brunel wrought iron design.

40. BI: 22 June 1836 I.K. Brunel to Grisell & Peto 25 June 1836 (Adrian Vaughan Collection)
41. Levien, D.L. 'The First Contract ever let by the Great Western Railway Company', *Great Western Railway Magazine* September 1835 pp 481-483.
42. See also: Coleman, T. *The Railway Navvies* (London: Penguin Books, 1970).
43. BI: DM 1298/2/71 Letter from Hammond to Archibald 14 July 1837.
44. BI: DM1298/2/6 Letter to GWR from Thompson & Co. 19 January 1837.
45. BI: DM1298/2/38 Letter from Hammond to R. Archibald, 24 April 1837.
46. This stretch was not actually completed until 1840.
47. TNA: RAIL 250/1 Great Western Railway: Report of the Fifth Half-yearly General Meeting 27 February 1838.
48. TNA: RAIL1014/1 Great Western Railway: Specification of Rails May 1837.
49. Charles Richardson Diary: 29 November 1837–13 January 1838.
50. TNA: RAIL 250/1 Great Western Railway: Report of the Fifth Half-yearly General Meeting 27 February 1838.

Chapter 4

1. Simmons, J. (Ed) *The Birth of the Great Western Railway: Extracts from the Diary and Correspondence of George Henry Gibbs* (Bath: Adams & Dart, 1971) p 26.
2. *Extracts from the Diary and Correspondence of George Henry Gibbs* p 27.
3. BI: DM1758/15 Transcription of letter from IKB to Charles Saunders 3 December 1837.
4. TNA: RAIL1016/5 Great Western Railway: Second Half-Yearly General Meeting 25 August 1836.
5. Brunel, I. *The Life of Isambard Kingdom Brunel Civil Engineer* (London: Longmans, 1870)
6. For further detail, see: Arman, B. *The Broad Gauge Engines of the Great Western Railway: Part 1 1837–1840.*
7. Arman, B. *The Broad Gauge Engines of the Great Western Railway: Part 1 1837–1840* pp 15-64
8. See also: Platt, A. *The Life & Times of Daniel Gooch* (Stroud: Alan Sutton, 1987)
9. Burdett-Wilson, R.B. (Ed.) *Sir Daniel Gooch: Memoirs & Diary* (Newton Abbot: David & Charles 1972) p 27.
10. *Sir Daniel Gooch: Memoirs & Diary* p 32.

11. *Sir Daniel Gooch: Memoirs & Diary* p 28.
12. *Sir Daniel Gooch: Memoirs & Diary* p 32.
13. Bryan, T. *North Star: A Tale of Two Locomotives* (Swindon: Thamesdown Borough Council, 1989) pp 4-5.
14. *Morning Star* was not finally delivered to the GWR until 1839.
15. Clark, E.T. *Great Western Railway Guide Book* (London: Smith & Ebbs, 1838) p 15.
16. *Extracts from the Diary and Correspondence of George Henry Gibbs* p 34.
17. *Extracts from the Diary and Correspondence of George Henry Gibbs* p 35.
18. *Extracts from the Diary and Correspondence of George Henry Gibbs* p 32.
19. BI: DM1306/7/5 IKB to Thomas Guppy 11 May 1838.
20. BI: DM1758/15 Transcription of letter from IKB to Charles Saunders 3 December 1837.
21. 'The Opening of the Great Western Railway', *Great Western Railway Magazine* Vol XLVII September 1935 p 465.
22. The station was renamed Taplow in 1869 when a new station was built in Maidenhead itself.
23. An average of 70 coaches per day were said to run through Maidenhead before the opening of the GWR.
24. See: Karau, P, Clark, M & Wells, M. 'The Great Western Railway at Maidenhead' *British Railway Journal Special GWR Edition* 1985 pp 3-19.
25. 'The Great Western Railway', *The Times* 5 June 1838 p 5.
26. 'The Company's First Western Terminus – Maidenhead', *Great Western Railway Magazine* September 1935 p 467.
27. *Extracts from the Diary and Correspondence of George Henry Gibbs* p 40.
28. *Extracts from the Diary and Correspondence of George Henry Gibbs* pp 43-44.
29. TNA: RAIL1016/5 Great Western Railway: Sixth Half-Yearly General Meeting August 15, 1838.
30. A detailed report of the meeting was given in a Special Supplement to *The Bristol Mercury* published on 18 August 1838.
31. *Extracts from the Diary and Correspondence of George Henry Gibbs* p 54.
32. *Extracts from the Diary and Correspondence of George Henry Gibbs* p 61.
33. *Sir Daniel Gooch: Memoirs & Diary* p 38.
34. *The Railway Times* 5 January 1839.

35. Great Western Railway: Report of a Special General Meeting of the Company 9 January 1839.
36. *The Railway Times* January 9 1839.
37. *The Railway Times* 19 January 1838.
38. *Sir Daniel Gooch: Memoirs & Diary* p 35.
39. Reproduced in MacDermot pp 30-31
40. Measom, G. *The Illustrated Guide to the Great Western Railway* 1852 Reprinted Edition (Reading: Berkshire Country Library, 1985) pp 18-19.
41. TNA: RAIL1149/2 p64. Letter from IKB to William Payne 29 August 1836.
42. TNA: RAIL250/82 p116 Letter from IKB to GWR Directors 19 June 1838.
43. Quoted in *How the Great Western Came to Berkshire* (Reading: Berkshire County Library, 1985) pp 13-14.
44. *The Berkshire Chronicle* 11 January 1839.
45. *Extracts from the Diary and Correspondence of George Henry Gibbs* p 68.
46. Measom, G. *The Illustrated Guide to the Great Western Railway* 1852 p 28.

Chapter 5

1. Casson makes the reasonable assumption that the C&GWUR was backed by those in Cheltenham who wished to maintain its status as a spa town in competition with Bath, but he offers no evidence for this. Casson, M. *The World's First Railway System: Enterprise, Competition, and Regulation on the Railway Network in Victorian Britain* (Oxford: Oxford University Press, 2009) p. 114.
2. Despite strong opposition from the London & Birmingham Railway that was promoting a competing route to London 20 miles shorter. See MacDermot, E.T. *History of the Great Western Railway Volume 1 1833-1863* p 79.
3. TNA: RAIL 1016/5 Great Western Railway Second Half-Yearly Meeting 25 August 1836.
4. STEAM:2015/51/3 Oxford & Great Western Union Railway Proofs 1837-1838.
5. STEAM:2015/51/3 Evidence of William Hemming.
6. STEAM:2015/51/3 Petition of Chancellor, Masters and Scholars of the University of Oxford, 9 June 183.8
7. STEAM:2015/51/3 Petition of the inhabitants of the City of Oxford and its immediate neighbourhood, June 1838.
8. McDermot, E.T. *History of the Great Western Railway Volume 1 1833–1863* pp 94-95.

9. See: Clifford, D. *Isambard Kingdom Brunel: The Construction of the Great Western Railway* (Reading: Finial Publishing, 2006) pp 213-217.
10. See Vaughan, A. *A Pictorial Record of Great Western Architecture* (Oxford: OPC, 1977) p 196.
11. Clark, E.T. Great *Western Railway Guide Book* (London, Smith & Ebbs 1838) p 18.
12. Transcript of Letter from W. Taylor, Didcot 12 June 1839. Great Western Trust Didcot Collection (Ref: GWT/2014/P7)
13. TNA: RAIL 1016/5 Great Western Railway: Report of the Tenth Half-Yearly General Meeting 24 August 1840.
14. Quoted in: McDermot, E.T. *History of the Great Western Railway Volume 1 1833–1863* p 60.
15. *The Great Western, Cheltenham and Great Western and Bristol and Exeter Railway Guides* (London: James Wyld, 1839) p 111.
16. BI: DM162/8/1/3/GWR Sketchbook 11/folio 13 Hay Lane Plans.
17. The canal proved not to be the best source of water and its supply to the works was always problematic. This was not properly solved by the GWR until the establishment of a borehole at Kemble, which supplied the works by pipeline in 1906.
18. TNA RAIL 1008/82 Letter from Daniel Gooch to I.K. Brunel 13 September 1840.
19. *Sir Daniel Gooch: Memoirs & Diary* p 40.
20. TNA: RAIL 1016/5 Great Western Railway: Report of the Eleventh Half-Yearly General Meeting 25 February 1841
21. See: Cattell, J. & Falconer, K. *Swindon: The Legacy of a Railway Town* (London: HMSO, 1995) pp 16-17.
22. See also: Arman, B. *The Broad Gauge Locomotives of the Great Western Railway Part 3: 1846-1852* (Lydney: Lightmoor Press, 2022) pp 35-40.
23. The Fawcett List is reproduced in: Peck, A *The Great Western at Swindon Works* (Oxford: OPC, 1983) pp 276-277.
24. The GWR Locomotive Factory *The Illustrated Exhibitor and Magazine of Art*. 1852.
25. Quoted in Rolt, L.T.C. *Isambard Kingdom Brunel* p 140.
26. TNA: RAIL252/174 Swindon Refreshment Rooms Lease 18 December 1841.
27. Quoted in: Rolt, L.T.C. *Isambard Kingdom Brunel* p 140. I have been unable to trace the original letter from which these famous comments are quoted.
28. *The Great Western, Cheltenham and Great Western and Bristol and Exeter Railway Guides* (London: James Wyld, 1839) p 111.
29. For more detail on the early inhabitants of New Swindon see: Cockbill, T. *A Drift of Steam: The Founders of Modern Swindon* (Swindon: Quill Press 1992).

30. Quoted in: Cattell, J. & Falconer, K. *Swindon: The Legacy of a Railway Town* (Swindon: RCHME 1995).
31. Which included Swindon from 1782.
32. TNA: RAIL 1016/5 Great Western Railway Twelfth Half-Yearly Meeting 23 August 1841.
33. Bourne, J.C. *The History and Description of the Great Western Railway* (London: David Bogue, 1846) p 49.
34. See: Pugsley, A. (Ed) *The Works of Isambard Kingdom Brunel* (London: Institution of Civil Engineers, 1976) Chapter 2 Tunnels pp 40-48.
35. For further reports of accidents in the tunnel see: Swift, A. *The Ringing Grooves of Change* (Bath: Akeman Press, 2006) pp 226- 235.
36. Report from the Select Committee on Railway Labourers. House of Commons 28 July 1846.
37. 'Great Western Railway – The Box Tunnel', *The Railway Times* 1840 pp 548-549.
38. Swift, A. *The Ringing Grooves of Change* p 226.
39. BI: Transcribed letter from I.K. Brunel 23 January 1839 (Adrian Vaughan Collection).
40. *The Great Western, Cheltenham and Great Western and Bristol and Exeter Railway Guides* p 118.
41. Brunel, I. *The Life of Isambard Kingdom Brunel Civil Engineer* p 83.
42. Clifford, D. *The Construction of the Great Western Railway* (Reading: Finial Publishing, 2006) pp 65-66.
43. Bourne, J.C. *The History and Description of the Great Western Railway* pp 52-53.
44. Brunel, I. *The Life of Isambard Kingdom Brunel Civil Engineer* p 83.
45. For a detailed analysis of Brunel's tunnel portal designs and inspiration see: Pragnell, H. 'Tunnels in Arcadia: Isambard Kingdom Brunel's Portal Designs for the Great Western Railway', *Architectural History* 63 (2020) pp 143-169.
46. *The Railway Times* 4 June 1841.

Chapter 6

1. Arthurton, A.W. 'The Principal Stations of the Great Western Railway No.6 – Bristol (Temple Meads)' *Great Western Railway Magazine* December 1907. P 264.
2. See: Harris, P. *Bristol's Railway Mania:1862-1864* (Bristol: Bristol Branch of the Historical Association, 1987).
3. BI: DM/10/2a/folio 46-48 IKB to Daniel Gooch, 2 January 1841.
4. See: Binding, J. *Brunel's Bristol Temple Meads* (Hersham: OPC, 1991) p 27.

5. Bourne, J.C *The History and Description of the Great Western Railway* (London: David Bogue, 1846) p 55.
6. TNA RAIL1008/62 Great Western Railway: Reports of Bristol Sub-committee 1835-1841.
7. TNA RAIL1008/62 Great Western Railway: Reports of Bristol Sub-committee 1835-1841.
8. Simmons, J. (Ed) *The Birth of the Great Western Railway: Extracts from the Diary and Correspondence of George Henry Gibbs* (Bath: Adams & Dart, 1971) p 69.
9. Measom, G. *The Illustrated Guide to the Great Western Railway* 1852 p 10.
10. Bourne, J.C. *The History and Description of the Great Western Railway* p 55.
11. Pevsner, N. *The Buildings of England: North Somerset and Bristol* (London: Penguin Books, 1958) p 421.
12. Pugin, A.W. *An Apology for The Revival of Christian Architecture* (London: John Weale, 1843) p 11.
13. Binding, J. *Brunel's Bristol Temple Meads* (Hersham: OPC, 2001) p 43.
14. Great Western Railway: Progress of Works Sub-Committee 15 July 1840.
15. BI: PLB IKB Letterbook 18 April 1841.
16. So named because of the clock mounted on the left-hand entrance gateway.
17. Oakley, M. *Bristol Temple Meads – 180 Years 1840-2020* (Bristol: Mike Oakley, 2020)
18. Pevsner, N. *The Buildings of England: North Somerset and Bristol* p 421.
19. GWR business continued to be conducted at the station until 1855, when overall administration was moved to Paddington.
20. Jenkins, S. *Britain's 100 Best Railway Stations* (London: Penguin Viking, 2017) p144.
21. BI:DM 1758/1
22. Bristol & Exeter Railway: Prospectus, Bristol, 21 October 1835 (STEAM Collection)
23. BI: PLB1 p 502.
24. No office appointment diary for 1835 survives, so it is difficult to judge how much time Brunel spent on the B & ER Survey.
25. Bristol & Exeter Railway: First General Meeting of the Proprietors 2 July 1836.
26. *The Great Western, Cheltenham and Great Western and Bristol and Exeter Railway Guides* (London: James Wyld, 1839) p 237.
27. Bristol & Exeter Railway. Act of Parliament 1836.

28. Bristol had been badly affected by cholera in 1832, and the disease was not totally eradicated. It returned in 1849. See: Hardiman, S. *The 1832 Cholera Epidemic and its Effect on Bristol* (Bristol: Bristol Branch of the Historical Association, 2005).
29. TNA: RAIL 75/49 Directors Report to Bristol & Exeter General Meeting 31 August 1837.
30. Latimer, J. *The Annals of Bristol: Volume 3, Nineteenth Century* (Reprinted by Bristol: Kingsmead Press, 1970) p 222.
31. TNA: RAIL75/1 Copy of Agreement signed at Reading on 14 August 1840 between the deputies of the Great Western and Bristol and Exeter Railway Companies.
32. MacDermot, E.T. 'The Bristol & Exeter Railway', *Great Western Railway (Bristol) Lecture & Debating Society* 21 March 1923.
33. BI: PLB 1 I.K. Brunel to Gravatt, 15 April 1859.
34. *Somerset Country Gazette* 16 November 1839.
35. BI: PLB 2a Brunel to Gravatt 23 July 1840.
36. BI PLB 2a Brunel to Gravatt 4 August 1840.
37. BI: PLB 2a Brunel to Gravatt 6 April 1841.
38. BI: PLB 2a Brunel to Gravatt 18 June 1841 (transcription).
39. For a detailed and scholarly analysis of both the relationship between Brunel and Gravatt and the construction of the Bristol & Exeter Railway see: Greenfield, D. 'I.K. Brunel and William Gravatt, 1826–1841: Their professional and personal relationship', PhD Thesis, University of Portsmouth, March 2011.
40. BI: DM1758/5/4/1 & 2 IK Brunel Appointment Diary for 1842.
41. Williams, M. 'When Brunel Swallowed a Half-Sovereign' *Brunel Society Gazette*, September 1980 pp 2-7.
42. Elton, J. 'Brunel in Clevedon' *Commemorative brochure for Clevedon Cavalcade of Cars* (Clevedon: *The Clevedon Mercury*, 2006) pp 4-5.
43. TNA: RAIL75/7 Bristol & Exeter Board Minutes 3 September 1842.
44. TNA RAIL 75/51 Bristol & Exeter Railway, Mr Brunel's Report: Bristol 21 August 1844.
45. Latimer, T. *Annals of Bristol* (Bristol Kingsmead Press Reprint, 1971) pp 222-223.
46. Quoted in: Maggs, C. *The Bristol & Gloucester Railway* (Lingfield, Oakwood Press, 1969) p 12.
47. The fact that northbound Bristol and Gloucester passengers needed to change trains at Gloucester does not seem to have been mentioned.
48. Cheltenham & Great Western Union Railway Act 1836.
49. BI: DM162/10/2/Folio 27: Letter from I.K. Brunel estimating the cost of the line from Cheltenham to near Swindon and Gloucester and Cirencester branches. 14 April 1836.

50. See also: Griffin, P. 'Charles Richardson, Civil Engineer, in Gloucestershire 1835–1845', *Gloucestershire Society for Industrial Archaeology Journal* 2010. pp 28-38.
51. Charles Richardson Diary 1 August 1837.
52. The embankment suffered further slips in 1841, one set of tracks being put out of action. The other, a traveller observed, appeared 'to be hanging by a thread'. See: McDermot. E.T. *History of the Great Western Railway Volume 1 1833-1863* pp 83-84.
53. MacDermot, E.T. *History of the Great Western Railway Volume 1 1833-1863* p 84.
54. See: Popplewell, L. *A Gazetteer of the Railway Contractors of The West Country 1830–1914* (Ferndown: Megdellen Press, 1983).
55. Charles Richardson Diary, May, June and July 1837.
56. See: Maggs, C. *The Swindon to Gloucester Line* (Stroud: Alan Sutton, 1991) p 17.
57. See: Household, H. 'Sapperton Tunnel, Western Region', *The Railway Magazine* February 1950 pp 79-82.
58. See: Lewis, B. *Brunel's Timber Bridges and Viaducts* (Hersham: Ian Allan, 2007) pp 49-54.
59. Lewis, B. *Brunel's Timber Bridges and Viaducts* p 49.

Chapter 7

1. Most of the items in the dinner service are now displayed in the Being Brunel museum at the SS *Great Britain* in Bristol. The centrepiece was melted down to raise funds for a memorial window to the engineer installed in Westminster Abbey in 1868.
2. *Cheltenham Examiner* 25 June 1845. Sekon erroneously gives the date as 1846.
3. See Appendix 2.
4. The definitive work on the subject is: Lewin, H.G. *The Railway Mania and its Aftermath 1845–1852* (London: The Railway Gazette 1936).
5. BI: I.K. Brunel to Charles Saunders 29 October 1845 (Adrian Vaughan Collection).
6. Quoted in: Vaughan, A. *Isambard Kingdom Brunel: Engineering Knight Errant* (London: John Murray, 1991) p 219.
7. The Great Western would also clash with the London & South Western Railway as it promoted railways south of its Bristol to London main line.
8. Quoted in MacDermot, E.T. *History of the Great Western Railway Volume 1 1833-1863* p 108.
9. MacDermot, E.T. *History of the Great Western Railway Volume 1 1833-1863* p 109.

10. STEAM Collection: *A Railway Travellers Reasons for Adopting Uniformity of Gauge Addressed to I.K. Brunel* Esq. (London: Charles Cundell, 1846).
11. *A Letter to the Directors of the Great Western Railway Shewing the public evils and troubles attendant upon their Break of Gauge and Pointing Out the Remedy by an Old Carrier* (Manchester: Bradshaw & Blacklock, June 1846).
12. Contemporary writers tended to use the terms 'broad' and 'narrow gauge' although we would now call the latter 'standard gauge' following the results of the Gauge Commission.
13. TNA: RAIL558/990 *Oxford, Worcester & Wolverhampton Railway Prospectus*, September 1844.
14. The story of the Oxford & Rugby Railway is complicated by rivalry and intrigue between the Grand Junction Railway and the GWR. A good summary can be found in: Jenkins, S.C. & Quayle, H.I. *The Oxford, Worcester & Wolverhampton Railway* (Blandford: Oakwood Press, 1977) pp 12-15.
15. Quoted in: Christiansen, R. *A Regional History of the Railways of Great Britain: Volume 7 The West Midlands* (Newton Abbot: David & Charles, 1973) p 65.
16. BI: DM1758/5/6/1 IK Brunel Appointment Diary for 1844.
17. TNA: RAIL1149/28 Report of the Commissioners of Railways 19 April 1847 p 12.
18. Sekon, G.A. *A History of the Great Western Railway* (London: Digby Long & Co, 1895) p 118.
19. *Oxford, Worcester and Wolverhampton and Oxford and Rugby Railway Bills: Statement of the Promoters* 1845, accessed digitally from the British Library.
20. Quoted in: Sekon, G.A. *A History of the Great Western Railway* (London: Digby Long & Co, 1895) p 199.
21. *The Railway Times* 26 July 1845.
22. *The Railway Times* 9 July 1845.
23. *The Railway Times* 12 July 1845.
24. TNA: RAIL250/775 *Report of the Gauge Commissioners and the Observations Thereon* 1846. The full report of the Gauge Commissioners extends to more than 850 pages of evidence and conclusions, so for brevity I have included quotations drawn from Sidney, S. *Extracts from the Gauge Evidence 1845 and the History and Prospects of the Railway System* (London: Edmonds, 1845) Reprinted 1971 by SR Publishers Ltd.
25. Sidney, S. *Extracts from the Gauge Evidence 1845 and the History and Prospects of the Railway System* pp 127-128.

26. Sidney, S *Extracts from the Gauge Evidence 1845 and the History and Prospects of the Railway System* p 216
27. Burdett-Wilson, R.B. (Ed.) *Sir Daniel Gooch: Memoirs & Diary* (Newton Abbot: David & Charles 1972) p 49.
28. Sidney, S. *Extracts from the Gauge Evidence 1845 and the History and Prospects of the Railway System* p 353. One wonders how serious Brunel was being when he made this statement.
29. Vaughan, A. *Isambard Kingdom Brunel: Engineering Knight Errant* p 219.
30. *The Railway Times* 21 February 1846.
31. It was estimated that it would cost more than £1,000,000 to convert the whole British railway network to broad gauge. See: *Bradshaw's Railway Gazette* 7 April 1846.
32. *Sir Daniel Gooch: Memoirs & Diary* p 52.
33. *Sir Daniel Gooch: Memoirs & Diary* p52.
34. Brunel, I. *The Life of Isambard Kingdom Brunel Civil Engineer* (London: Longmans, 1870) p 119.
35. Sekon, G.A. *A History of the Great Western Railway* p 141.
36. Brunel, I. *The Life of Isambard Kingdom Brunel Civil Engineer* p 120.
37. A large number of these tracts are listed in: Ottley, G. (Ed.) *A Bibliography of British Railway History* (London: HMSO, 1983) pp 362-365.
38. STEAM Collection: *A Railway Traveller's Reasons for Adopting Uniformity of Gauge Addressed to I.K. Brunel* Esq. (London: Charles Cundell, 1846).
39. *A Letter to the Directors of the Great Western Railway Shewing the public evils and troubles attendant upon their Break of Gauge and Pointing Out the Remedy by an Old Carrier* (Manchester: Bradshaw & Blacklock, June 1846).
40. 'L.s.d.'. *The Broad Gauge: the bane of the Great Western Railway, with an account of the present and prospective liabilities saddled on the proprietors by the promoters of that peculiar crotchet.* (London, 1846).
41. See also: Fara, P. 'Great Debates: The Battle of the Gauges', *History Today* March 2022 pp 85-89.
42. For an analysis of how the GWR recovered from the broad gauge, see: Bryan, T. *The Golden Age of the GWR 1892–1914* (Peterborough: PSL Books, 1991).
43. *The Exeter Flying Post* 25 May 1843, quoted in *Brunel's Teignmouth*, Teignmouth & Shaldon Museum Monograph No. 14 2006 p 14.
44. BI:DM1758/5/5 IK Brunel Appointment Diary for 1843.
45. TNA RAIL 1110/421 South Devon Railway Prospectus, 1844.

46. Brunel's Office diary is frustratingly blank for the early part of 1843, so his attendance at the committee stages of the SDR bill is not recorded.
47. Quoted in: *Brunel's Teignmouth*, Teignmouth & Shaldon Museum Monograph No. 14 2006 p 18.
48. Nicknamed 'Mr Punch's line', this was the West London Line, a short railway used by Samuda to test the atmospheric idea.
49. Stephenson, R. *Report on the Atmospheric Railway System* (London: John Weale, 1844) p 40.
50. TNA: RAIL631/44 Atmospheric Railway: I.K. Brunel's Reports & Correspondence.
51. *Sir Daniel Gooch: Memoirs & Diary* p 46.
52. See: Hadfield, C *Atmospheric Railways: A Victorian Venture in Silent Speed* (Newton Abbot: David & Charles, 1967).
53. See also: Buchanan, A. 'The Atmospheric Railway of I.K. Brunel', *Social Studies of Science* Vol 22 May 1992 pp 231-243.
54. The 'Northern Leviathan' referred to by Gill was George Stephenson.
55. *The Railway Times* 14 March 1845.
56. The London & Croydon Railway designed by William Cubitt began trials running with an atmospheric system in August 1845.
57. BI: *Minutes of the Evidence of the Commission on Atmospheric Railways* 1845 (Adrian Vaughan Collection) p 36.
58. Brunel is once again being economical with the truth as the Croydon line had been beset with problems during this period and had closed for a period in July 1846.
59. *The Railway Times* 5 September 1846.
60. Adrian Vaughan suggested that Brunel was forced to attend.
61. Brunel, I. *The Life of Isambard Kingdom Brunel Civil Engineer* p 165.
62. 'The South Devon Railway', *The Times* 31 October 1846.
63. BI: *Portbury Pier & Railway Company Report* April 1845.
64. *The Bristol Mercury* 4 September 1847.

Chapter 8

1. Quoted in: Awdry, C. *Brunel's Broad Gauge Railway* (Oxford: OPC, 1992) p 64.
2. An Act for regulating the Gauge of railways 18 August 1846 Vic Cap LVII.
3. *Railway Times* 18 October 1845.
4. TNA: RAIL588/1388 *Narrative of Circumstances Connected with Oxford, Wolverhampton & Wolverhampton Railway*. GWR April 1854.
5. Quoted in: Jenkins, S.C. & Quayle, H.I. *The Oxford, Worcester & Wolverhampton Railway* (Oakwood Press, 1977) p 17.

6. TNA: RAIL588/1388 *Narrative of Circumstances connected with Oxford, Wolverhampton & Wolverhampton Railway*. GWR April 1854.
7. *Railway Times* 21 March 1846.
8. *Railway Times* 3 October 1846.
9. TNA: RAIL588/1388 *Narrative of Circumstances Connected with Oxford ,Wolverhampton & Wolverhampton Railway*. GWR April 1854: Letter from I.K. Brunel to Francis Rufford, Chairman of the OW&WR, 8 September 1848.
10. BI: IK Brunel Journal 1 May 1849. Transcription: Adrian Vaughan Collection.
11. For further detail on OW&WR contracts and opening dates see: Popplewell, L. (Ed) *A Gazetteer of the Railway Contractors and Engineers of Central England 1830–1914* (Bournemouth: Melledgen Press, 1986).
12. RAIL588/1391 *Condensed Narrative of Circumstances Connected with Oxford, Wolverhampton & Wolverhampton Railway* p 21.
13. TNA: RAIL 1008/35 f.79. 8/3/52 (Copy: BI Collection)
14. RAIL588/1391 *Condensed Narrative of Circumstances Connected with Oxford, Wolverhampton & Wolverhampton Railway* p 29.
15. A drawing of a standard OW&WR station is shown on p 31 of: Jenkins, S.C. & Quayle, H.I. *The Oxford, Worcester & Wolverhampton Railway* (Oakwood Press, 1977)
16. Biddle, G. *Britain's Historic Railway Buildings: An Oxford Gazetteer of Structures and Sites* (Oxford: Oxford University Press, 2003) p 304-305 and Listed Grade II – https://historicengland.org.uk/listing/the-list/list-entry/1053261.
17. For a full list of timber bridges and viaducts see: Lewis, B. *Brunel's Timber Bridges and Viaducts* (Hersham: Ian Allan Publishing, 2007)
18. Mickleton Tunnel Contract Correspondence between IKB and W. Williams (Contractor) 1850–1851, Great Western Trust Collection GWT/1985 G49.3.
19. The English spelling. In Welsh it is known as Porthdinllaen.
20. *London, Worcester, and Porth-Dynllaen Railway*: handbill (undated) Great Western Trust Collection.
21. *Herepath's Journal* December 1845 Quoted In: Jones, S.K. *Brunel in South Wales: Volume 2: Communications and Coal* (Stroud: History Press, 2006).
22. BI: Brunel Appointment Diary for 1836 DM1758/5/3.
23. *Railway Times* 9 September 1843.
24. The full story of the SWR can be found in Stephen K. Jones' definitive and scholarly *Brunel in South Wales: Volume 2: Communications and Coal* (Stroud: History Press, 2006).

25. RAIL1014/2/38 South Wales Railway Prospectus 1844.
26. In view of Brunel's wish to avoid Gloucester in his revised SWR route, the decision not to use the original Gloucester & South Wales Railway name may be significant.
27. *Railway Times* 9 January 1845.
28. For further detail see: Jones, S.K. *Brunel in South Wales: Volume 2: Communications and Coal* p 79-81.
29. 8 & 9 Vic Cap CXC, 'An Act for the Making of a Railway to be Called the "South Wales Railway"', 4 August 1845. Work on the Monmouth branch did not progress past some initial earthworks and the scheme had been abandoned by 1849.
30. 'The South Wales Railway Bill', *Evening Mail* 27 June 1845.
31. Spelled 'Fowy' in the Act.
32. TNA: RAIL1110/431 South Wales Railway Company General Meeting 26 August 1846.
33. In March 1854 the SWR Chairman was able to report that the line would be completed within budget, noting that they had spent £3,500,000 with only seven miles of railway to completed. *Hereford Times* 11 March 1854.
34. Reports had circulated in the South Wales press in 1844 that atmospheric traction might be used.
35. Bradshaw's 1850, p. 191, Bradshaw's 1851, p. 221.
36. 10 Vic Cap, 'An Act for extending the line of the South Wales Railway, and for making certain alterations of said Railway, and certain branch railways in connexion therewith', 27 July 1846.
37. See Chapter 7 and Daniel Gooch about Brunel's departure to the continent on the eve of the Gauge Commission.
38. TNA: RAIL1110/431 South Wales Railway: Company General Meeting 26 August 1846.
39. TNA: RAIL1110/431 South Wales Railway: Mr Brunel's Report February 22 1847.
40. TNA: RAIL1110/431 South Wales Railway: Special General Meeting 10 April 1847.
41. TNA: RAIL1110/431 South Wales Railway: Engineer's Report August 19 1848.
42. Lewis, B. *Brunel's Timber Bridges and Viaducts* (Hersham: Ian Allan Publishing, 2007) pp 68-78.
43. *Monmouthshire Beacon* 8 June 1850.
44. *Monmouthshire Beacon* 8 June 1850.
45. *Hereford Journal* 28 June 1850.
46. *Brunel's Tubular Suspension Bridge over the River Wye 1856* (Chepstow: Chepstow Society Reprint, 1976) p 6.

47. *The Cork Constitution* Tuesday 13 April 1852.
48. Brunel's Wye Bridge at Chepstow has not survived. Weakened by heavy traffic during the Second World War, despite repairs it was replaced after ninety years in 1962.
49. See: C. Price, M. Fishguard, 'Abermawr, Neyland: Building the broad gauge in Pembrokeshire', *Journal of the Railway & Canal Historical Society* Volume 40 No.242 November 2021 pp 326-342.
50. TNA: RAIL1110/431 South Wales Railway: Half-Yearly Meeting August 1851.
51. The Eastern Steam Navigation Company was established in 1851. The SS *Great Eastern* was moored at Neyland for two extended periods.
52. See Bryan, T. *The Golden Age of the GWR 1892–1914* (Peterborough: PSL Books, 1991).
53. TNA: RAIL1110/43: Great Western Railway: Special Meeting, November 1846.
54. TNA: RAIL 704 Vale of Neath Railway: Minutes of meeting held on 21 May 1845.
55. See: Jones, S.K. *Brunel in South Wales: Volume 2: Communications and Coal* pp 178-179 and Wilson, A.N. *The Victorians* pp 224-225.
56. Quoted in: Jones, G.B. & Dunstone, D. *The Vale of Neath Line* (Gomer Press: Llandysul, 1999) p 27.
57. Quoted in: Richards, S. *The Vale of Neath Railway* (Cardiff: Morgannwg, 1978) p 5.
58. BI: DM1758/5/13/folio 15 IK Brunel Appointment Diary for 1850.
59. *Hereford Times* 11 March 1854.

Chapter 9

1. In late 1847 the Great Western Railway had laid off more than 1,000 of the 1,800 workers at Swindon as the value of Company shares tumbled. Brunel contributed £100 to a fund to help workers and their families. See: Peck, A.S. *The Great Western at Swindon Works* (Oxford: Oxford Publishing Company, 1983) p 43.
2. BI: DM1281/1 PLB 23 August 1848. The underlining is Brunel's.
3. Brunel's ship was not salvaged from the beach where it had run aground until September 1847. The Great Western Steamship Company were bankrupted by the disaster and both the SS *Great Britain* and his first ship, the SS *Great Western*, were subsequently sold.
4. For further detail on Brunel's Watcombe Estate, see: Tudor, G. *Brunel's Hidden Kingdom* (Paignton: CMP, 2007). The Watcombe notebooks are held at the Brunel Institute. Despite much planning, the house was never constructed during Brunel's lifetime.

5. These railways included the Oxford, Worcester & Wolverhampton, Oxford & Rugby, South Wales and Vale of Neath described in previous chapters.
6. See Appendix 1.
7. Brunel, I. *The Life of Isambard Kingdom Brunel Civil Engineer* (London: Longmans, 1870) p 91.
8. It is estimated that Brunel made 220 appearances in Parliament during his career, more than any engineer of his generation. See Gren, A. *The Foundation of Brunel's Great Western Railway* (Kettering: Silver Link Publishing, 2003).
9. For detail on the first Paddington station see: Tutton, M. *Paddington Station 1833–1854* (Railway & Canal Historical Society, 1999).
10. MacDermot, E.T. *History of the Great Western Railway Volume 1 1833-1863* (London: Great Western Railway, 1927, Revised Edition Ed. C.R. Clinker Ian Allan, 1964) p 333.
11. TNA RAIL1005/82 Transcript of Half-Yearly Meetings.
12. Dobson's train shed roof at Newcastle completed in 1849-50 was probably the first significant structure of this type.
13. See: Gloag, J. & Bridgwater, D. *A History of Cast Iron in Architecture* (London: Allen & Unwin, 1948) pp 199-202.
14. Given that the structure was intended as a temporary exhibition hall, these criticisms seem harsh.
15. See: Thorne, R. 'Masters of Building 7: Paddington Station', *Architects Journal* 13 November 1985.
16. Later renamed Praed Street.
17. Tutton, M. *Paddington Station 1833–1854* (Railway & Canal Historical Society, 1999) p 4.
18. TNA: RAIL258/361 Letter from J. Burke to Charles Saunders 4 September 1850.
19. TNA: RAIL250/4 Great Western Railway Board Minutes 19 December 1850.
20. TNA: RAIL1008/35 I.K. Brunel to Charles Saunders 17 March 1852.
21. For a detailed analysis of the history and design of Paddington see: Brindle, S. *Paddington Station: Its History and Architecture* (Swindon: English Heritage, 2004).
22. Quoted in Rolt, L.T.C p189.
23. The BW&DR Act was passed on 22 July 1847 but the GWR absorbed the B&OJR by Act of 31 August 1848 and opened on 1 October 1852.
24. Burdett-Wilson, R.B. (Ed.) *Sir Daniel Gooch: Memoirs & Diary* (Newton Abbot: David & Charles 1972) p 62.

25. Wolverhampton remained the northernmost outpost of Brunel's broad gauge.
26. For detailed background see: Hale, M *Brunel's Broad Gauge in The Black Country* (Dudley: Woodsetton Monograph, 1997)
27. Quoted in: MacDermot, E.T. *History of the Great Western Railway Volume 1 1833-1863* p 143.
28. Reported in *The Railway Times* 28 September 1839.
29. Quoted in: Lewis, B. *Brunel's Timber Bridges and Viaducts* p 66.
30. Simmons, J. *The Railway In Town & Country 1830–1914* (Newton Abbot: David & Charles, 1986) p 308. Also: Biddle, p 67
31. Wilts, Somerset & Weymouth Railway – First General Meeting *The Railway Times* 11 October 1845.
32. TNA: RAIL1014/1 Handbill: Railway To Devizes – 'Notice of a Public Meeting to consider the expediency of applying to the Court of Queen's Bench for a Mandamus to enforce the Completion of the Branch railway to this town', 26 March 1852.
33. Biddle, *Britain's Historic Railway Buildings* p168.
34. Quoted in: Phillips, D. *Steaming Through the Cheddar Valley* (Oxford: OPC 2012) p 9.
35. The Bodmin & Wadebridge Railway opened in 1834 was both the first to use steam traction and to carry passengers in Cornwall.
36. *The Illustrated Times* Saturday 19 September 1857, p 198.
37. See: Cross-Rudkin, P. & Chrimes, M. (Eds) *Biographical Dictionary of Civil Engineers Vol 2 1830–1890* (London: Institution of Civil Engineers, 2008) pp 550-552.
38. *The Railway Times* 12 December 1846.
39. The line would have had nine tunnels, eight miles of cuttings and nine miles of embankments. See: Woodfin, R.J. *The Cornwall Railway* (Truro: Bradford Barton, 1972) p 2.
40. TNA: RAIL1066/719 Cornwall Railway: Mr Brunel's Evidence before the Committee of the House of Commons. 2 June 1845.
41. TNA: RAIL 134/1 Cornwall Railway Resolution 25 July 1845.
42. *The Railway Times* 6 September 1845.
43. The GWR subscribed £75,000, the Bristol & Exeter £112,000 and the South Devon £150,000.
44. Vic Cap CCCXXXV 'An Act for making a Railway and other Works from Plymouth to Falmouth, and other Places in the Country of Cornwall to be called the Cornwall Railway', 3 August 1846.
45. The proposals for branch lines to Falmouth and St Ives were rejected.
46. Quoted in: Jenkins, S.C. & Langley, R.C. *The West Cornwall Railway: Truro to Penzance* (London: Oakwood Press, 2002) p 37.

47. Quoted in: Jenkins, S.C. & Langley, R.C. *The West Cornwall Railway: Truro to Penzance* (London: Oakwood Press, 2002) p 33.
48. Simmons, J. 'The Railway in Cornwall 1835–1914', *Journal of the Royal Institution of Cornwall.* Vol IX. Quoted in: Binding, J. *Brunel's Cornish Viaducts* (Penrhyn: Atlantic Publishing, 1993.) p 11.
49. TNA: RAIL 134/17 Cornwall Railway: A collection of letters and report from IK Brunel relating to the construction of the line and the building of Saltash Bridge 1853–1859 (transcribed and typed by the GWR). Letter to E.H. Bond, 13 June 1853.
50. For the story of the SS *Great Eastern* see: Emmerson, G.S. *The Greatest Iron Ship: SS Great Eastern* (Newton Abbot: David & Charles, 1980). When shipping magnate and MP William Schaw Lindsay was asked by IKB regarding the *Great Eastern*, 'How will she pay?' he replied, 'Turn her into a show...Send her to Brighton, dig out a hole in the beach and bed her stern in it...her hold would make magnificent saltwater baths.' Lindsay noted: 'Brunel never forgave me.' Quoted in Lindsay, B. *William Schaw Lindsay: Victorian Entrepreneur* (Stroud: Amberley books 2023) p 159.
51. RAIL/134/17 Report to Directors of the Cornwall Railway 2 August 1855.
52. Brunel built few bridges in cast iron. See: Mylius,' A Cast Iron Excuse', *New Civil Engineer* 24 February 2005.
53. See: Binding, J. *Brunel's Cornish Viaducts* (Penrhyn: Atlantic Publishing, 1993) and Woodfin, R.J. *The Cornwall Railway* (Truro: Bradford Barton, 1972) pp 44-58.
54. RAIL/134/17 Report to Directors of the Cornwall Railway 24 May 1854.
55. In a final irony, one set of chains had came from the original Clifton Bridge.
56. Cross-Rudkin, P. & Chrimes, M. (Eds) *Biographical Dictionary of Civil Engineers Vol 2 1830–1890* (London: Institution of Civil Engineers, 2008) p125-126.
57. BI: DM326/24 IK Brunel to Charles Saunders 30 August 1854.
58. RAIL/134/17 Brunel to C. Gainsford Esq. 28 April 1857.
59. *History of the Royal Albert Bridge* (Devonport: Wood & Tozer, 1859)
60. The official party from Truro missed the festivities as the locomotive hauling their train failed near Liskeard.
61. For the complete story of the Saltash Bridge see: Binding, J. *Brunel's Royal Albert Bridge* (Truro: Twelveheads Press 1997).
62. RAIL/134/17 Report to Directors of the Cornwall Railway 18 August 1859.

Epilogue: After Brunel

1. Quoted in Binding, J. *Brunel's Royal Albert Bridge* (Truro: Twelveheads Press 1997) p118.
2. TNA: RAIL1014/30 R.P. Brereton 28 September 1859.
3. The Briton Ferry Trust is working to preserve and enhance the surviving dock. See: www.brunelquays.co.uk
4. *The Cambrian* 23 August 1861.
5. 'Southampton' was dropped from the name in the final scheme.
6. BI: PLB DM162/10/4 IK Brunel to Charles Richardson 14 September 1858.

Appendix 1

1. Other openings included The Oxford Branch (1844), Windsor Branch (1849), The Wycombe Railway (1854), Uxbridge Branch (1856, Henley Branch (1857) and the Brentford Branch (1859)
2. The complicated history of this railway included separate openings for the Thingley Junction – Westbury (1848) Westbury – Frome (1850), Frome – Radstock and Frome -Yeovil (1856), Bradford on Avon to Bathampton (1857) and Devizes Branch (1857).

Appendix 2

1. BI: DM162/10/2a folio 80-83 PLB: IK Brunel to Edwin Gower 28 August 1841.
2. Buchanan, R.A. *The Overseas Projects of I K Brunel* (Bath: Centre for the History of Technology, Science and Society, 1983) p 3.
3. BI: DM1758/5/13/folio 83 IK Brunel Appointment Diary for 1851.
4. Brunel, I. *The Life of Isambard Kingdom Brunel Civil Engineer* (London: Longmans, 1870) p 91.

Sources and Bibliography

Much of the primary material used in this book was drawn from the collections held at the Brunel Institute at the SS Great Britain in Bristol, which is a collaboration between the SS Great Britain Trust and the University of Bristol. I also consulted the Brunel Collection and Great Western Railway files held at the National Archive at Kew. Other original material was sourced from the British Library at St Pancras, the Search Engine at the National Railway Museum at York, the library at the STEAM Museum in Swindon, the Great Western Trust at Didcot, the Senate House Library, University College of London and the local collections of Bristol and Swindon Central Libraries. Local newspaper reports were accessed either from the libraries noted above, or from the British Newspaper Digital Archive.

Individual file references are recorded in the notes section and the following secondary sources were also consulted:

Ackroyd, P. *Dominion* (London: Picador, 2018)

Acworth, W.M. *The Railways of England* (London: John Murray, 1890)

Arman, B. *The Broad Gauge Engines of the Great Western Railway Part 1: 1837-1840* (Lydney: Lightmoor Press 2018)

Arman, B. *The Broad Gauge Engines of the Great Western Railway Part 2: 1840-1845 The Broad Gauge Engines of the Great Western Railway Part 2: 1840-1845* (Lydney: Lightmoor Press 2020)

Arman, B. *The Broad Gauge Engines of the Great Western Railway Part 3: 1846-1852* (Lydney: Lightmoor Press 2022)

Awdry, C. *Encyclopaedia of British Railway Companies* (London: Guild Publishing, 1990)

Awdry, C. *Brunel's Broad Gauge Railway* (Oxford: OPC, 1992)

Barrie, D.S.M. *A Regional History of the Railways of Great Britain: Vol.12: South Wales* (Newton Abbot: David& Charles, 1980)
Barrie, D.S.M. The *Taff Vale Railway* (Oakwood Press, 1950)
Bates, S. *Penny Loaves & Butter Cheap: Britain in 1846* (London: Head of Zeus, 2014)
Biddle, G. *Britain's Historic Railway Buildings: An Oxford Gazetteer of Structures and Sites* (Oxford: Oxford University Press, 2003)
Biddle, G. *The Railway Surveyors* (London: Ian Allan, 1990)
Binding, J. *Brunel's Bristol Temple Meads* (Hersham: OPC, 2001)
Binding, J. *Brunel's Cornish Viaducts* (Penrhyn: Atlantic Publishing, 1993)
Binding, J. *Brunel's Royal Albert Bridge* (Truro: Twelveheads Press 1997)
Bourne, J.C. *The History and Description of the Great Western Railway* (London: David Bogue, 1846)
Brindle, S. *Brunel: The Man who Built the World* (London: Weidenfield & Nicholson, 2006)
Brindle, S. *Paddington Station: Its History and Architecture* (Swindon: English Heritage, 2004)
Briwnant-Jones, G. & Dunstone, D. *The Vale of Neath Line* (Llandyssul: Gomer Press, 1996)
Brunel, I. *The Life of Isambard Kingdom Brunel Civil Engineer* (London: Longmans, 1870)
Bryan, T. *Broad Gauge Railways* (Oxford: Shire Books, 2012)
Bryan, T. *Brunel: The Great Engineer* (Hersham: Ian Allan, 1999)
Buchanan, R.A. *Brunel* (London: Hambeldon & London, 2002)
Buchanan, R.A. *Brunel in Bath* (Bath: Holbourne Museum, 2006)
Buchanan, R.A. *The Engineers: A History of the Engineering Profession in Britain 1750–1914* (London: Kingsley, 1989)
Buchanan, R.A. *The Overseas Projects of I K Brunel* (Bath: Centre for the History of Technology, Science and Society, 1983)
Buchanan, A. & Williams, M. *Brunel's Bristol* (Revised edition Bristol: Redcliffe Press, 2005)
Burdett-Wilson, R.B. (Ed.) *Sir Daniel Gooch: Memoirs & Diary* (Newton Abbot: David & Charles 1972)
Cannadine, D. *Victorious Century: The United Kingdom 1800–1906* (London: Penguin, 2018)
Cattell, J. & Falconer, K. *Swindon: The Legacy of a Railway Town* (Swindon: RCHME 1995)
Christiansen, R. *A Regional History of the Railways of Great Britain, Vol. 13, Thames and Severn* (Newton Abbott: David & Charles, 1981)
Clifford, D. *The Construction of the Great Western Railway* (Reading: Finial Publishing, 2006)

Clinker, C.R. (intro), *Bradshaw's Railway Manual Shareholders' Guide and Directory 1869*, 2nd edn. (Newton Abbott: David & Charles Reprints, 1969).

Conder, F.R. *Personal Recollections of English Engineers* (London: Hodder & Stoughton, 1868)

Corfield, P.J. *The Georgians* (London: Yale University Press, 2022)

Cross-Rudkin, P. & Chrimes, M. (eds) *Biographical Dictionary of Civil Engineers Vol 2 1830–1890* (London: Institution of Civil Engineers, 2008)

Day, L. *Broad Gauge* (London: HMSO, 1984)

Doe, H. *The First Atlantic Liner: Brunel's Great Western Steamship* (Stroud: Amberley Press, 2017)

Edwards, J. (Comp) *Bradshaw's General Railway Directory, Shareholders' Guide, Manual and Almanack, Etc* (London: W. J. Adams, and Manchester: Bradshaw & Blacklock, 1851)

Ferriday, P. (Ed) *Victorian Architecture* (London Jonathon Cape, 1963)

Francis, J. *A History of the English Railway: Its Social Relations & Revelations 1820–1845* (London: Longman Brown, Green & Longmans, 1851)

Freeman, M. *Railways and the Victorian Imagination* (London: Yale University Press, 1999)

Garnsworthy, P. *Brunel's Atmospheric Railway* (Broad Gauge Society, 2013)

Grant, D.J. *Directory of the Railway Companies of Great Britain* (Kibworth Beauchamp: Matador, 2017).

Great Western Railway *The Great Western Railway Centenary 1835–1935* (London: GWR, 1935)

Gregory, R.H. *The South Devon Railway* (Oakwood Press, 1982)

Gren, A. *The Foundation of Brunel's Great Western Railway* (Kettering: Silver Link Publishing, 2003)

Hale, M. *Brunel's Broad Gauge in The Black Country* (Dudley: Woodsetton Monograph, 1997)

Hale, M. *The Oxford, Worcester & Wolverhampton Railway Through the Black Country* (Dudley: Woodsetton Monograph, 1995)

Hayes, D. *The First Railways: An Atlas of Early Railways* (London: Harper Collins, 2017)

Jenkins, S.C. & Quayle, H.I. *The Oxford, Worcester & Wolverhampton Railway* (Oakwood Press, 1977)

Jenkins, S.C. & Langley, R.C. *The West Cornwall Railway: Truro to Penzance* (London: Oakwood Press, 2002)

Jones, S.K. *Brunel in South Wales: Volume 1 In Trevithick's Tracks* (Stroud: History Press, 2005)

Jones, S.K. *Brunel in South Wales: Volume 2: Communications and Coal* (Stroud: History Press, 2006)

Jones, S.K. *Brunel in South Wales: Volume 3: Links with Leviathans* (Stroud: History Press, 2009)

Kentley, E. (Ed) *Isambard Kingdom Brunel: Recent Works* (London: Design Museum 2000)

Latimer, T. *Annals of Bristol in the Nineteenth Century* (Bristol: W.F. Morgan, 1887

Lewin, H.G. *The Railway Mania and its Aftermath 1845–1852* (London: The Railway Gazette, 1936)

Lewis, B. *Brunel's Timber Bridges and Viaducts* (Hersham: Ian Allan Publishing, 2007)

Lindsay, B. *William Schaw Lindsay: Victorian Entrepreneur* (Stroud: Amberley Books 2023)

MacDermot, E.T. *History of the Great Western Railway Volume 1 1833-1863* (London: Great Western Railway, 1927, Revised Edition Ed. C.R. Clinker, Ian Allan, 1964)

Maggs, C. *The Bristol & Gloucester Railway* (Oakwood Press, 1969)

Maggs, C. *The GWR Bristol to Bath Line* (Stroud: Sutton Publish Company 2001)

Maggs, C. *The Swindon to Gloucester Line* (Stroud: Alan Sutton Publishing, 1991)

Phillips, D. *How the Railway Came to Berkshire: A Railway History 1833-1852* (Reading: Berkshire County Library, 1985)

Peck, A. *The Great Western at Swindon Works* (Oxford: Oxford Publishing Company, 1983)

Platt, A. *The Life & Times of Daniel Gooch* (Stroud: Alan Sutton, 1987)

Popplewell, L. (Ed) *A Gazetteer of the Railway Contractors and Engineers of Central England 1830–1914* (Bournemouth: Melledgen Press, 1986)

Popplewell, L. (Ed) *A Gazetteer of the Railway Contractors and Engineers of Wales and the Borders 1830–1914* (Bournemouth: Melledgen Press, 1984)

Popplewell, L. (Ed) *A Gazetteer of the Railway Contractors and Engineers of the West Country 1830–1914* (Bournemouth: Melledgen Press, 1983)

Pugsley, A. (Ed) *The Works of Isambard Kingdom Brunel* (London: Institution of Civil Engineers & University of Bristol, 1976)

Railway Correspondence and Travel Society. *The Locomotives of the Great Western Railway: Part Two Broad Gauge* (RCTS, 1952)

Rolt, L.T.C. *Isambard Kingdom Brunel* (London: Longmans, 1957)

Sekon, G.A. *A History of the Great Western Railway* (London: Digby Long & Co, 1895)

Simmons, J. (Ed) *The Birth of the Great Western Railway: Extracts from the Diary and Correspondence of George Henry Gibbs* (Bath: Adams & Dart, 1971)

Simmons, J. *The Railway in Town & Country 1830-1914* (Newton Abbot: David & Charles, 1986)
Simmons, J. *The Victorian Railway* (London: Thames & Hudson, 1991)
Simmons, J. & Biddle, G. (Ed) *The Oxford Companion to British Railway History* (Oxford: Oxford University Press, 1997)
Smiles, S. *Lives of the Engineers* (London: Folio Society Edition 2006)
St. John-Thomas, D. *A Regional History of the Railways of Great Britain: Vol.1 The West Country*. (Newton Abbot: David & Charles, 1988)
Stroud Local History Society. *The Railway Comes to Stroud 1845* (Stroud: SLHS, 1995)
Swift, A. *The Ringing Grooves of Change: The Coming of the Railway to Bath* (Akeman Press, 2006)
Tutton, M. *Paddington Station 1833–1854* (Railway & Canal Historical Society, 1999)
Vaughan, A. *A Pictorial Record of Great Western Architecture* (Oxford: Oxford Publishing Company, 1971)
Vaughan, A. *The Intemperate Engineer: Isambard Kingdom Brunel in his own Words* (Hersham: Ian Allan Publishing, 2010)
Vaughan, A. *Isambard Kingdom Brunel: Engineering Knight-Errant* (London: John Murray, 1991)
Whishaw, F. *The Railways of Great Britain and Ireland* (London: John Weale, 1842)
Williams, A. *Brunel & After* (London: Great Western Railway, 1925)
Wilson, A.N. *The Victorians* (London: Hutchinson, 2002)
Woodfin, R.J. *The Cornwall Railway* (Truro: Bradford Barton, 1972)

Illustrations

Index